Mirror Earth

The Georgian Star
Echo of the Big Bang
Other Worlds
The Light at the Edge of the Universe

Mirror Earth

The Search for Our Planet's Twin

Michael D. Lemonick

Walker & Company

New York

Published by Walker Publishing Company, Inc., New York
A Division of Bloomsbury Publishing

All papers used by Walker & Company are natural, recyclable
products made from wood grown in well-managed forests.
The manufacturing processes conform to the environmental
regulations of the country of origin.

LIBRARY OF CONGRESS CATALOGING-IN-PUBLICATION DATA

Lemonick, Michael D., 1953–
Mirror Earth : the search for our planet's twin /
Michael D. Lemonick.—1st U.S. ed.
p. cm.
Includes bibliographical references and index.
ISBN: 978-0-8027-7900-7 (hardback)
1. Extrasolar planets. 2. Earth. 3. Planetology. I. Title.
QB820.L46 2012
523.2'4—dc23
2012009787

Visit Walker & Company's website at www.walkerbooks.com

First U.S. edition 2012

1 3 5 7 9 10 8 6 4 2

Typeset by Westchester Book Group
Printed in the U.S.A. by Quad/Graphics, Fairfield, Pennsylvania

For Seymour Lemonick,
whose love and support
have been there since
before I can remember

CONTENTS

INTRODUCTION

"THE EARTH GOES around the Sun. *That's* what's going
on!"

These were the last words my father ever said to me. He
said them—almost shouted them—with a vehemence and
conviction that startled me. The fact that he could say anything at all was itself a surprise. He'd been semicomatose for
nearly a week, unable to initiate movement or speech on his
own. He would sometimes respond to questions, but only in a
hoarse whisper, so faint that it was impossible to understand
what he was saying. That afternoon, I'd arrived at his room at
the nursing wing of the retirement community he lived in,
and said, "Dad, what's going on?" with the artificial hearti-
ness you sometimes put on to convince yourself and others that
everything is perfectly normal, even though it's anything but
normal. He barked out that single phrase, and then didn't speak
after that. Two days later, he died.

At the time, those final words didn't make a bit of sense to
me, but when I began to write this book, I thought about them
again, and remembered the stories my father used to tell me

when I was very young. He was a professor of physics at Princeton University and an extraordinarily popular teacher. When they hear my last name, gray-haired men still tell me how much they loved taking his classes a half century ago or more. He would present long-established ideas in physics—Newton's laws of gravity or Maxwell's equations of electromagnetism or Einstein's theories of relativity—as though he had just stumbled on these mind-blowing truths for the first time. He could barely contain his excitement, and his students couldn't help sharing it. The fact that the Earth goes around the Sun was a big deal when Copernicus first proclaimed it in the 1500s. And it still is!

I never took one of those classes. I got a different sort of physics education, delivered in the form of stories my father would tell late at night, as we drove home from visiting his family or my mother's in Philadelphia. He told stories about atoms and molecules and planets and stars, pitched at a level an eight-year-old could understand, but filled with the same excitement and wonder he shared with his students at Princeton. I remember the time he told the story of Halley's Comet— about how it returned every seventy-six years to light up the night sky, how it was passing by the year Mark Twain was born and again as he lay on his deathbed. It would be coming again, my father promised, in 1986, when I would be an unimaginable thirty-three years old. I couldn't wait.

Those late-night stories about the universe didn't inspire me to become a physicist—too much math! They did, however, inspire me to become a journalist who never wanted to write about anything but science. I was intrigued, not only by what astronomers had learned already, but also by mysteries still unanswered. When my father first told me about the cosmos, astronomers didn't know about quasars, or black holes, or pulsars. They didn't know that most of the matter in the universe is not the atoms that make up stars, planets, and people, but rather a mysterious, invisible substance known as dark matter. They knew that the universe was expanding, but had no idea that the expansion is accelerating, driven by an equally mysterious force known as dark energy.

And they didn't know the answer to perhaps the oldest questions of all: Do planets orbit distant stars? Do any of them harbor life? Is the human species alone in the universe?

It wasn't that astronomers and physicists hadn't tried to answer these questions, but planets around other stars are excruciatingly hard to see. They lie tens of trillions, or even hundreds of trillions, of miles away. Stars appear tiny when you look up into the night sky, but planets are far tinier, and far dimmer. Finding planets around even the nearest stars turned out to be so difficult that hunting for distant worlds had become a fringe area of astronomy. Anyone who thought he or she could figure out how to do such a thing shouldn't be taken very seriously.

All of that changed in 1995, however, when Swiss astronomers found the very first planet orbiting a Sun-like star. In 1996, American astronomers found several more. The Americans, especially, had staked their careers on finding planets (the Swiss were working on other things as well), and they hadn't gotten a lot of respect. But once this small handful of pioneers had shown it was possible to find planets, their colleagues quickly changed their attitudes, and started looking too. Over the next decade and a half, the number of known worlds beyond our own solar system would climb into the tens, then the hundreds, then to more than a thousand.

But the ultimate goal of finding a Mirror Earth—a planet of about the same size as our home planet, with the right mix of land and ocean and temperature so that life might have established a foothold and gone on to thrive—remained just out of reach.

It won't be out of reach for long, though. By early 2012, it was clear that a Mirror Earth was finally within astronomers' grasp. It likely would be only months, rather than years, before astronomers would be able to take my father's dying words one step further by declaring: "A Mirror Earth goes around a Sun-like star—and here's where it is!"

Five billion years ago, the Milky Way didn't look much different than it does today. It was, and remains, an enormous pinwheel some five hundred quadrillion miles across, made up of hundreds of billions of stars, rotating once every two hundred million years or so. Between the stars lay immense, swirling clouds of gas and dust, and every so often, one of these clouds would begin to collapse under its own gravity. The collapse might be triggered by a shock wave from a supernova—a star ending its life in a titanic explosion—or it might be caused by the blast of radiation from a hot, blue supergiant star, or simply by the gravity of a star lumbering by.

Once it started, the cloud kept falling in on itself, getting smaller and denser, spinning faster and faster and, because of the spin, flattening itself out like a pancake. When it was about five billion miles across, one one thousandth of its original size, and one hundred million miles or so thick, the cloud was dense enough and spinning so fast that gravity could no longer force it to shrink any further. Particles of iron, nickel, and silicates—molecules made of oxygen and silicon—collided and stuck together and began to grow larger.

But by now, the intense pressure in the densest part of the cloud—the very core—was heating it up to temperatures of millions of degrees. The searing heat ripped electrons away from atoms and forced atomic nuclei to overcome their natural repulsion, releasing enormous energy. The core of the pancake burst into life as a newly formed star—a gigantic, self-perpetuating thermonuclear furnace that would burn for the next ten billion years. The star would one day be known as the Sun. Once the Sun flared into life, it had a quick and profound effect on the pancake that still swirled around it. The heat and light and intense magnetic fields emanating from the young star would collectively have swept most of the remaining gases out toward the edges of the pancake, leaving the rocky and metallic solids behind.

The grains grew into pebbles, and on up in size until they were what planetary scientists call planetesimals—rocky, metallic objects big enough for their gravity to pull them into a roughly spherical shape. The largest asteroids that remain in the solar system, including Ceres and Vesta, resemble those early planetesimals. In those early days, however, there were hundreds of thousands of them whipping around the Sun, slamming together, breaking apart, re-forming, and ultimately growing larger and larger, with fewer and fewer planetesimals to wreak havoc on one another. Close to the Sun, the chaos would end with four major rocky planets: Mercury, Venus, Earth, and Mars. (Earth's relatively huge Moon, scientists believe, was created when one final Mars-size object slammed into Earth and ripped off a chunk).

Rocky planets formed in the outer solar system as well. They were up to ten times as massive as the Earth, with ten times the gravity, and they began to suck in gas—mostly hydrogen, but also an enormous amount of water vapor and nitrogen and carbon compounds—forming the thick, deep atmospheres of present-day Jupiter, Saturn, Uranus, and Neptune. The stuff that was left over, out beyond the orbit of Neptune, formed into the comets.

This is the picture astronomers put together during the 1960s, '70s, and '80s to explain how the solar system formed, and while they weren't certain of every step of the process, they were pretty confident that it all held together. That being the case, it made sense that other solar systems, if they existed, would form in more or less the same way. There would be local variations, of course, from one solar system to the next. The formation process couldn't have unfolded in *exactly* the same way every time. But just as every normally developed human is born with two legs at the bottom, two arms sticking out the sides, and a head on top, astronomers expected most solar systems to have rocky planets close in and gas-giant planets farther out.

Then came 1995. In the fall of that year, a Swiss observer named Michel Mayor announced that he'd discovered a planet orbiting a Sun-like star in the constellation Pegasus. This was a very big deal: It was the first such planet ever found, after thousands of years of speculation about whether such worlds might exist, and after decades of searching by prominent astronomers. But the planet was all wrong. It was impossible. It was a world more massive than Jupiter, but sitting so absurdly close to its star that a full orbit—a year, from the planet's point of view—took just four days. (In our solar system, Mercury, the planet closest to the Sun, takes eighty-eight days to go once around).

That first planet might have been some sort of bizarre fluke, but over the next few years, more and more of what astronomers were now calling "hot Jupiters" were found, all over the sky. The conventional wisdom about how solar systems form was clearly wrong, or at least incomplete. Theorists had made the classic error of thinking that the single example they had to work with was a typical example.

Observers, however, were thrilled. Until that first planet was found, many doubted that existing telescopes were powerful enough to find alien worlds. Now that observers knew otherwise, they switched by the score from whatever area of astronomy they'd been working in. They began searching for planets, bringing new ideas and energy to what had until then been a backwater of astronomy. And while these first solar systems were weird and crazy, they brought a renewed hope that Earth-like worlds, too, existed out among the stars, places where life, and maybe even intelligent life, had taken hold.

To qualify as a Mirror Earth, a planet presumably had to be more or less the size of the original—big enough to hold on to an atmosphere, but not so big that the atmosphere would be crushingly dense. It had to be made mostly of rock, like Earth: A planet that was the right size but mostly made of water or gas presumably wouldn't be hospitable. It had to orbit within the so-called habitable zone surrounding its star, the region where heat from the star was just enough to let water exist in liquid form, rather than purely as ice or water vapor. That's because water is a necessary ingredient for all known forms of life on Earth, and the same rule would presumably apply to a Mirror Earth as well.

Finding a world that met all these requirements wouldn't happen in a single stroke—in part because the features that distinguish a Mirror Earth from some other kind of planet require a different sort of observation, and in part because the technologies and the search strategies that would ultimately lead to that discovery had to be invented and tested and perfected and funded over periods that lasted years. Nobody could really predict when that discovery would finally happen; all they could do was press to find smaller and smaller worlds; figure out ways to estimate their composition, and sample their atmospheres, and calculate whether they were common or rare in the Milky Way.

In 1995, the idea of finding a Mirror Earth was still just a dream. Today, thanks to a dramatic series of advances in both technology and in the ways astronomers are conducting their searches, they're on the verge of doing just that. This book describes how it all unfolded.

Chapter 1

THE MAN WHO LOOKED
FOR BLINKING STARS

BILL BORUCKI WAS starting to look a little impatient. He had just hosted a press conference at which he and several other astronomers had talked about the latest results from NASA's Kepler space telescope. The event was over, and now it was time for lunch. There was really no time to lose. The press conference was held as part of the February 2011 meeting of the American Association for the Advancement of Science—the biggest general-science meeting of the year, hosted by the biggest scientific society in the world. It was taking place at the Washington, D.C., convention center, which consists of two huge concrete buildings straddling M Street, just north of the capital's main tourist area.

The convention center is so big that the cavernous basement of one of the buildings can hold major sporting events, and while some of the most eminent scientists in the world were on the upper floors pondering the deepest mysteries of the universe, a regional volleyball tournament was happening down below. Five hundred or so teenagers had been playing at least a dozen side-by-side, simultaneous matches, rotating on and off the courts in a continuous stream since early that morning.

Now, five hundred ravenous young people, along with parents, friends, and coaches, were streaming out to the dining area, where their warm-up suits and flushed cheeks contrasted sharply with the rumpled suits and relatively pasty faces of graduate students, professors, and research scientists, including not a few Nobel Prize winners. Neither group would have a wonderful experience (a review on Yelp.com read, in part, "One Starbucks for a whole convention, you're crazy. Hence the Disneyland-esque line we waited in . . . The food court is also pretty bad. A cabbage salad with chicken, ewww . . .").

Borucki might not have been aware of this impending dining disaster, but his impatience had a more immediate and obvious explanation. Seth Borenstein, the lead science reporter for the Associated Press, had missed the press conference, and he needed to catch up. "Sorry, no story today" are words that would never issue from Borenstein's lips. Unfortunately, there wasn't any actual news. It's not that the Kepler Mission itself wasn't newsworthy. Since its launch in the spring of 2009, Kepler has utterly revolutionized the search for planets circling stars beyond the Sun. Before the mid-1990s, the search had limped along for decades without success. Since the first planet was discovered in 1995, planet hunters had identified between four hundred and five hundred of these alien worlds (they're technically known as "extrasolar planets," or simply "exoplanets"), one by painstaking one. The number isn't exact because not everyone agrees that all of the discoveries are real.

Then Kepler came along. It was clear from the start that if the spacecraft worked as it was supposed to, it would blow the competition out of the water. This was true for several reasons. To start with, Kepler orbits high above Earth's atmosphere, like the Hubble Space Telescope, but even farther out in deep space. As a result, the atmosphere's blurring effect—the same thing that makes the stars seem to twinkle—isn't a problem. Another huge advantage is that Kepler doesn't look at one star at a time. It looks at more than 156,000 of them all at once, in a patch of the northern sky that lies between the constellations Lyra and Cygnus. And it keeps looking at those same 156,000 stars continuously, around the clock, day after day, month after month, year after year. That's impossible from the ground: When the Sun rises, the stars vanish. Since one night's observing isn't nearly enough to find a planet around a distant star, you have to keep returning to a given star many times, over many nights, to get any information worth using. Kepler doesn't return to any of its 156,000 stars because it never leaves them in the first place.

If Kepler were a general-purpose telescope—even one like the Hubble—it wouldn't be permitted to linger on a single patch of sky indefinitely. Most telescopes are used to study all sorts of cosmic phenomena, from distant galaxies to exploding stars to black holes. If you're looking for planets with a general-purpose telescope, you might get to use it for a few nights at most before the next astronomer in line gets her turn. If you're using the Hubble, which is vastly oversubscribed, you get more like a few hours. Kepler, by contrast, was built to look at only one tiny patch of sky for its entire working lifetime. It will never avert its gaze from these 156,000 stars.

By the time Borenstein, the AP reporter, had cornered Borucki, Kepler had proven itself to be technologically perfect, or pretty close to it. It had been staring at its target stars for nearly two years. Only the first six months' worth of observations had been fully processed, though. That's how long it takes for the Kepler team's computers to pore through the terabytes of electronic data beamed down from the spacecraft, letting custom-written algorithms flag the tiny changes in starlight that might (or might not) betray the presence of a planet, weeding out false positives—things that look like planets but aren't. If a potential planet passes all these tests, that's still not good enough. The software has to pick up the planet's signal not once, not twice, but at least three separate times for it to make the cut. Usually, though, since the signal is often very faint, it takes a lot more than three sightings: some run into the hundreds. And even then, a dozen or so mission scientists look at each of what they call KOIs—Kepler Objects of Interest. These are something like the persons of interest law-enforcement types talk about in criminal investigations. They're not being charged . . . yet. But you shouldn't be at all surprised if they end up being indicted.

The reason Borucki hadn't announced any new results at the Washington press conference was that he'd already presented everything he had just two weeks earlier, at a press conference at NASA headquarters, in Washington. There was plenty to say: In just the first six months' worth of observations, Kepler had come up with no fewer than 1,235 possible planets, about 90 percent of which were almost certainly real. Kepler had barely warmed up, and it had identified at least twice as many planets as all the astronomers in the world had found in the previous sixteen years. "Astronomers have cracked the Milky Way like a piñata," Dennis Overbye wrote in the *New York Times*, "and planets are now pouring out so fast that they do not know what to do with them all."

After he gave reporters the number of new planets—or, rather, "planet candidates," in the very careful language Kepler scientists prefer to use—Borucki explained, like a pollster projecting the outcome of an election based on just a small sample of voters, that Kepler is looking at only about 1/400th of the sky. If the spacecraft had been able to monitor the whole, he said (or if NASA had sent up four hundred identical Keplers, pointing in all different directions), they'd be talking not about twelve hundred planets, but about the more than four hundred thousand the probe would undoubtedly have seen.

It was a terrific story—two weeks earlier. The talk at the American Association for the Advancement of Science meeting was pretty much just a replay. For a science reporter—especially a reporter for the Associated Press, where late-breaking news is a specialty—that just wasn't good enough. Borenstein couldn't write a story that said, in essence, "The Kepler results announced two weeks ago are still true." So, while the editor of *Discover* magazine and I stood by, looking on in comradely amusement, Borenstein kept pushing the Kepler team leader to say something new. Borucki was clearly reluctant to be pushed.

"Okay, so if I understand you correctly," the reporter asked, "you said you've found 1,235 planet candidates, right?"

"That's right," said Borucki. He'd said this two weeks earlier. Borenstein knew it. But like a prosecutor in a courtroom, Borenstein was building his case.

"And of those, fifty-four are in the habitable zone of their stars?"

"Yes, that's correct."

Again, this news was two weeks old, but it was really important. The habitable zone is the orbital band surrounding a star where the temperature allows water to exist as a liquid rather than as ice or vapor. It's sometimes called the Goldilocks Zone, even by astronomers—even on NASA's website—since like the porridge in the fairy tale, it's not too hot and not too cold, but just right. Biologists have long insisted that water is essential for life, because nutrients can dissolve in it easily, to be distributed to every part of an organism. That's what blood does for most animals, and blood is mostly water. Life on Earth wouldn't be possible, says the conventional wisdom, if most of our water boiled off into the atmosphere or froze solid. Earth is inhabited because we live within the Sun's habitable zone.

If you're interested in finding life on other worlds—and that's what just about every scientist who hunts for planets is ultimately looking for—planets in the habitable zone are what you want. Planets about the size of Earth in the habitable zone are even better. The question of whether life exists beyond Earth is one of the oldest mysteries of nature, dating back at least to the ancient Greeks, and probably even further. At some times in history, the notion of alien life has been considered heretical; at others, learned men took it as a given that planets, both within and outside the solar system, were home to intelligent beings. The eighteenth-century astronomer William Herschel, who discovered the planet Uranus, was convinced that even the Sun was inhabited (he had a handy explanation for why the Sun creatures weren't incinerated).

Kepler isn't capable of answering the question of whether life exists on other worlds, but it can take the first step by finding an Earth-like planet, a Mirror Earth, where life could be thriving, at least in principle. Kepler was designed with several scientific objectives in mind, but number one on the list that appears on the mission website is this: "Determine the abundance of terrestrial [that is, Earth-size] and larger planets in or near the habitable zone of a wide variety of stars."

"So, as I understand it," continued Borenstein, pressing his interrogation, "there are about 300 billion stars in the Milky Way. If you're looking at 150,000 stars, and found 1,200 with planets, and 54 of those in the habitable zone . . . that means . . ." The reporter stared at the ceiling, wheels turning in his mind. Borucki looked on, politely. ". . . there should be something like 50 billion planets in our galaxy, right? And 500 million should be in the habitable zone."

Borucki thought about that for a moment. "Yes, that sounds right," he answered.

Two weeks earlier, Borucki had talked about a hypothetical four hundred thousand planets that could be detectable from Earth. Now, under intense, though friendly, questioning, he was admitting to five hundred million overall. Borenstein had his story. Later that day, the Knight Science Journalism Tracker, a blog that aggregates and reviews science stories, described it this way:

> Seth says 50 billion planets, minimum, in Milky Way. Nobody said that at the press conference. Minor consternation ensued among other reporters after he filed. How'd he get that angle? Explanation: Seth missed the press conference. Saw Borucki afterward talking with a few reporters including Michael Lemonick of *Time*. "Just a nice chat where you riff together," Borenstein says. Borucki says one in two stars has planets, Seth says let's do the math, Borucki complies and double checks, and that's why it can pay to be there in the flesh.

This remark about the flesh may well have been a dig at media outlets that are no longer willing to pay for reporters to go to conferences. The author of the blog post, Charles Petit, covered science for the *San Francisco Chronicle* for years, and takes a dim view of how his profession has been downsized. Still, the calculation done by the AP reporter was so simple and obvious that Borucki could easily have done it for the other reporters who were present, and for the hundreds more who get NASA press releases by e-mail. He could have done it for the press conference two weeks earlier.

The fact that he hadn't says a lot about Bill Borucki. Some astronomers are showmen—Neil deGrasse Tyson is a good example. Tyson, the director of the Hayden Planetarium in New York, is a serious and highly respected scientist, but he's also a frequent guest of both Jon Stewart and Stephen Colbert on Comedy Central. When he taught at Princeton (he had a part-time appointment to the faculty there for many years), his dynamic lectures drew students by the hundreds. Tyson is a tall, powerfully built man in his forties, a former athlete and dancer who once told me that "in high school, I was a nerd, but a nerd who could kick your butt." I once saw him at an astronomy conference striding across the hotel lobby in a form-fitting black workout outfit, complete with weightlifting gloves—a ninja astronomer heading for the gym. Every eye in the place followed him.

That makes Bill Borucki the anti-Tyson. He's in his early seventies, below average in height, with thinning hair and wire-rimmed glasses. He speaks softly where Tyson booms, and he pauses before he answers a question, where Tyson fires rapidly. Borucki wouldn't do well on *The Colbert Report*. His talks are generally delivered in a thoughtful, measured tone, without laugh lines or oratorical fireworks. Bill Borucki would look ridiculous in a black workout suit. Wearing a coat and tie, as he was when Borenstein cornered him after the Washington press conference, he could be mistaken for an accountant. In the more relaxed atmosphere of a recent astronomy conference, he ambled through the Washington State Convention Center in Seattle in a green windbreaker, looking like someone Hollywood might have cast as the clerk at an old-fashioned hardware store. Tyson grew up on New York's Upper West Side; he has an undergraduate degree in physics from Harvard and a Ph.D. from Columbia. Borucki grew up in Delavan, Wisconsin, in the space-happy 1950s. He built model

Bill Borucki (Courtesy of NASA)

rockets and, as head of the high school science club, organized the construction of a transmitter to contact UFOs. His undergraduate degree in physics comes from the University of Wisconsin, in Madison, and he has no Ph.D. at all—just a master's in physics, which he also got at Madison, in 1962.

Borucki's temperament is cautious enough that any calculation he might do for a press conference would tend to involve as little speculation as possible. When he came up with the figure of four hundred thousand planets across the entire sky, he was talking about planets that Kepler could in principle have detected if it was pointed in their direction. When Borenstein came up with five hundred million, he was talking about planets anywhere in the Milky Way, most of which couldn't be found by Kepler, or, for that matter, by any other telescope that could conceivably be built.

Beyond that, it would be easy for the general public to leap to premature conclusions about what Kepler had actually found. The fifty-four planets Borucki had announced in the habitable zones of their stars weren't necessarily anything like Earth. Many of them were much bigger—as big as Neptune or even Jupiter—and therefore not a place you could easily imagine finding life. Also, they didn't orbit stars like the Sun. They orbited red dwarf stars, which are significantly smaller and dimmer, and whose planets might not be good places for life to take hold. Sixty-eight Earth-size planets he'd also talked about, conversely, weren't in their stars' habitable zones, so there wasn't much chance of life there either.

In other words, the true story was a little complicated and a little subtle. There's often a tension between the reporter's need to have the most exciting story possible and the scientist's need to avoid too much hype. Borenstein walked that line like a tightrope. The story that came out of the Borucki-Borenstein encounter was absolutely accurate, but while it might have seemed otherwise to a lay reader, there was no new science in it—just a small victory for a reporter in his respectful but relentless tug-of-war with a scientist.

Talking to the press wasn't a problem Bill Borucki had to

deal with much in the early part of his career. His master's degree was enough to get him a job in the early sixties at the Ames Research Center, at Moffett Field, near Mountain View, California. His first assignment there: to help design heat shields to protect space capsules from burning up as they reentered Earth's atmosphere. The only breaking news on heat shields happened when Mission Control feared that the shield protecting John Glenn's *Friendship 7* capsule had worked its way loose during his first orbital flight, in 1962. If it had fallen off he would have been incinerated—but it wasn't loose after all. Even if it had fallen off, the heat shield itself wouldn't have been to blame.

During a visit to Ames in October 2010, I suggested to Borucki that the Kepler project was probably a lot more exciting than this first project must have been.

"Oh, my God," he said, his eyebrows rising with either dismay or pity, or maybe a little of both. "You don't know anything about heat shields, do you?"

"But . . . planets orbiting other stars," I protested feebly. "It's something the human race has dreamed of for thousands of years . . ."

"But imagine this reentry vehicle coming in," he protested back, "heating the shock wave in front of it to many thousands of degrees hotter than the surface of the Sun. That is interesting! Hotter than the surface of the Sun, and we've got to calculate the heat radiation on this shield—otherwise the astronauts die on their way back. That's a pretty impressive thing to work on." It was also a challenge: At the time, nobody had a very good idea about how to calculate what happens to air when you heat it up to tens of thousands of degrees. So Borucki and his colleagues began studying lightning, which does exactly that—taking images of the flashes and analyzing the light for evidence of what was happening to air molecules at these temperatures.

At the end of the 1960s, Borucki left the exciting world of heat shields. "After the day we were successful getting to the Moon," he said, "I moved over to the theoretical studies group at Ames." It wasn't as drastic a move as you might imagine, however, because he was still studying lightning. Except this time, the lightning was on Jupiter. Scientists using radio telescopes had detected bursts of static coming from the giant planet and suspected that lightning was the cause. Borucki and his colleagues wanted to understand how lightning on Jupiter might differ from lightning on Earth.

At first they were stuck with building laboratory experiments to simulate Jovian lightning, since at that point nobody had gotten a close look at the planet itself. They created miniature Jupiter atmospheres inside what amounted to huge test tubes. Then they fired lasers into them, which triggered electric sparks. And then they studied the flashes, just as they'd done with real lightning a few years earlier. In both cases, they had to build detectors that could measure changes in light with extraordinary precision. "You're trying to understand the fundamental measurements, you're trying to understand their time dependence, you're building photometers that people generally don't build that are running on the nanosecond level," said Borucki. When space probes finally detected the real thing during flybys of Jupiter in the late 1970s, the work

Borucki had done helped planetary scientists understand what was happening on the giant planet itself.

At about the same time Borucki was trying to understand lightning on Jupiter and the nature of the atmosphere that gave rise to it, he said, "there were seminars here at Ames. People would come and talk about future projects, going to Mars and finding life and things like that. Those were very inspirational, and I began to think about whether what I knew could help solve the problem of whether there's life in the galaxy."

His experience with model rocketry notwithstanding, Borucki didn't know much about going to Mars or designing experiments to look for life. He didn't know much about SETI, the Search for Extraterrestrial Intelligence, which his colleague Frank Drake had been working on since the early 1960s. Drake's idea was to listen with radio telescopes for the broadcasts, deliberate or inadvertent, that might be coming from alien civilizations. SETI is what Jodie Foster was doing in the movie *Contact*, although, naturally, with a little more drama and romance than the real thing.

In thinking about the search for life on other worlds, Drake had quickly realized that if alien civilizations really existed, they probably needed planets to live on. Not just any planets: They needed, as far as anyone knew, planets that were at least vaguely Earth-like, orbiting in their stars' habitable zones. In the 1960s nobody knew how many of these there might be in the Milky Way, if any at all. But Drake went ahead on the assumption that they must exist, given the vastness of the Milky Way. By the 1980s, astronomers still didn't know. They didn't even know if planets of any sort existed, Earth-like or not. A tiny handful of planets had been "discovered," starting in the 1960s, but every one of them had been undiscovered later on. The original detections had simply been mistakes.

Borucki thought maybe he could do better. He knew there was no way to see planets around other stars directly. Everyone knew they would be too faint, hidden in their star's much brighter glare. The astronomers who had made those earlier, false detections had used another method entirely. They'd looked at the stars themselves, hoping to see them wobble in place as the gravity of an orbiting planet tugged them just a tiny bit—first one way, then the other, as the planet circled from one side of the star to the other. This technique is known as astrometry, meaning "star measurement." The motion would be tiny, thus very difficult to see, because a planet is tiny compared with a star. It's not a huge surprise that those early discoveries were mistakes.

Partly as a result of these false detections, and partly because any sort of detection was impossible with any existing telescope, the search for planets around other stars had become a scientific backwater. There wasn't much point in looking until someone—NASA was the obvious choice—provided astronomers with powerful new instruments. If you can't see a tiny motion with an ordinary telescope, went the reasoning, build a gigantic telescope—or, since that's really expensive and impractical, use a group of smaller telescopes at one time to simulate a single, huge one. "People had sketched out systems that might do this," said Borucki, but it turned out to be a lot more difficult and expensive than anyone thought. (It wasn't until 2009 that two astronomers, using technology far more

sophisticated than anything available in the 1980s, finally announced they'd found the first planet ever discovered with
astrometry. NASA put out a press release with the headline
PLANET-HUNTING METHOD SUCCEEDS AT LAST. But it hadn't.
That discovery, like the earlier ones, is now widely considered
to have been a mistake.)

Borucki had no expertise in measuring the positions of stars
in the sky, but he did know how to measure light. "I said to
myself, 'Here's another way to do it. It's a rather simple way. It
doesn't require the kind of extreme equipment you need for
astrometry.'" So in the summer of 1983, he sat down with a
colleague named Audrey Summers and wrote a paper titled
"The Photometric Method of Detecting Other Planetary Systems."

The following year, their paper was published in the
planetary-science journal *Icarus*. The idea was simple enough:
If everything happened to be lined up in just the right way, a
planet orbiting a distant star would pass right between the star
and the Earth once a year. That's the planet's year, not ours,
since "year" really means "one orbit." A year on Venus lasts
224 days, so if an alien Bill Borucki were looking toward our
solar system from just the right angle and saw Venus passing
by, he'd see it a second time 224 days—one year—later. If he
spotted Mars, "one year later" would be 686 days; and for
Mercury, it would be 88 days.

If you were watching a star when one of these crossings
took place (they're officially known as "transits") you'd see the
star dim just a little, as the planet blocked some of the starlight. If you plotted the star's brightness over time, the dim-

ming would show up as a dip in an otherwise more or less steady glow. If it really were a transiting planet, the dip would repeat, like clockwork, every time the planet came around again on its orbit. The plot is known as a light curve; the regularity and appearance of the dip (V-shaped? U-shaped?) should tell the astronomers whether they were seeing a planet or something else.

That's what Borucki and Summers proposed to look for. They realized that you'd have to be looking at a distant solar system precisely edge-on to see this happen. Solar systems are tilted in random ways as seen from Earth, so only a fraction of them would be lined up the right way. In order to stand a chance of spotting even a single transit, therefore, you'd have to monitor hundreds, or even thousands, of stars at once. The easiest planets to see would obviously be the biggest ones, since they block the most light. But big planets like Jupiter orbit pretty far from their stars, as far as anyone knew at the time. Jupiter's year is about eleven Earth years long. So if your telescope was powerful enough to see only the dimming of a big planet, you'd have to wait a long time to see anything at all, even in an edge-on solar system. If you're looking at many hundreds of stars at once, a few score should be edge-on, and in some fraction of those, a Jupiter should be just about to make its transit. "Based on the stated assumptions," wrote Borucki and Summers, "a detection rate of one planet per year of observation appears possible."

This "simple way" did, however, require light detectors of unprecedented precision. If a Jupiter-size planet transited in front of a Sun-like star, the star's light should dim by about 1 percent, or one part in a hundred. At the time, when astronomers talked about high-precision photometry—that is to say, brightness measurements—they were talking about a sensitivity of one part in ten, or ten times less sensitive. "The first thing we had to do," said Borucki, "was to show we could build photometers with the necessary precision."

He knew it couldn't be done with photomultipliers, the detector technology astronomers were using at the time. Borucki thought it might be possible to use CCDs, or charge-coupled devices, a sort of detector-on-a-chip that had been invented in 1969 at Bell Laboratories. CCDs are the detectors that have replaced film in modern digital cameras. At the time, though, they were rare and expensive. Astronomers were just beginning to adopt them. "Younger astronomers were familiar with CCDs," he said, "but for the older ones, the attitude was, 'If God wanted you to have CCDs you would have been . . . born with a CCD in your mouth instead of a silver spoon or . . . something like that."

So Borucki and a couple of colleagues set up an experiment in the basement. "We built this thing with bricks and aluminum," he said, "and we had a light shining under an aluminum plate with a bunch of holes in it." Little holes were dim stars and big holes were bright stars. At first, the CCDs didn't seem to be sensitive enough to make the measurements he needed, but Borucki wrote a series of equations that corrected for the inaccuracies. "You can correct it out to ten parts, even one part per million," he said, "even with a poor detector." They'd shown, in other words, that they could find planets, and not just Jupiters but smaller planets as well—if someone would let them try. But when Borucki tried to sell the project, he met with a brick wall of resistance. "You go to all sorts of meetings," he said, "and you tell other astronomers, 'You know, we can find other planets. We can find small planets. We can find them with a CCD.'" The other astronomers would say no, that's impossible. "They would get up and show why it couldn't be done. I would say, 'We've done it. It can be done.' They would go to my boss and see if I could get taken off the project, because obviously we were wasting money, but he had enough faith in us that he let us continue."

His boss evidently had faith to spare. The CCD experiments happened in the late 1980s, and in the early 1990s Borucki took his sales pitch on the road. He went to NASA with an official proposal for a space-based planet-hunting telescope. It was rejected. He addressed the agency's criticisms, and reproposed the mission. He was rejected. In all, Kepler was rejected four or five different times before the satellite was finally approved in 2000. It launched in 2009, a quarter of a century after Borucki's first theoretical paper in *Icarus*. Natalie Batalha, who is now Borucki's deputy principal investigator on the Kepler Mission, was in high school when he wrote the original paper. Now she has the office next door to his, and I visited her the day after I saw him.

"It really takes a unique person to create a mission like this," she said. "Bill has this personality trait where negativity just rolls off of him. He doesn't accept it at all. I have never seen the guy take anything personally. You get rejected and they tell you, 'This is bad for the following reason,' and he doesn't take it the way most people would. He could get a major rejection in the mail one morning and in the afternoon still have the *cojones* to go to the administrators and ask for fifty thousand dollars to build whatever. Not many people can do that. It's incredible. It's been a pleasure to watch that and watch his positivity and persistence. It's been a life lesson for me. And it's a huge part of the Kepler story."

Chapter 2

THE MAN WHO LOOKED
FOR WOBBLING STARS

B<small>Y THE TIME</small> Bill Borucki was feeling the full force of
press and public attention weighing down on him in
2011, Geoff Marcy had already been dealing with it for fifteen
years. At this point, Marcy had been interviewed for literally
hundreds—probably thousands—of newspaper and magazine
articles, radio and TV shows, books and documentaries. Unlike Borucki, he seemed to relish all of it. Marcy invariably
comes across in these interviews as engaging, relaxed, funny.
He also comes across that way in person, at scientific lectures
and public talks. He's given these by the hundreds as well, and
they're hugely popular. He has a way of conveying the excitement of discovery and a sense of awe about the universe in an
impressively informal, intimate way. He seems impossible to
rattle. I watched him show up at a meeting of amateur geologists one night in the VFW hall in Orinda, California, only to
learn that there was no extension cord anywhere in the building (no venue seems to be beneath him, as long as the audience is genuinely interested). He couldn't project his slides on

the screen, so he put a chair on top of a table, propped up his laptop, and mesmerized everyone in the audience even though most of them couldn't see much of anything.

Marcy is like the best teacher you ever had—warm, engaging, enthusiastic, funny, and ridiculously knowledgeable. Granted, he's got an unfair advantage over some scientists in commanding your attention, given that his area of expertise isn't, say, dung beetles, or the history of sheet metal. But he does have a way of explaining the search for planets—and, for the past decade and a half, their discovery—in a way that anyone can understand and appreciate. For his students at the University of California, Berkeley, he really is the best teacher most of them have ever had. During a visit to his office, I watched him stop and talk to a group of students, joking with them but also asking about their research projects, about which he seemed fully up to speed. I thought they were grad students; a senior professor at a major university wouldn't normally have a lot of contact with undergrads, except in groups of one hundred or more in a large lecture room. The professor certainly wouldn't know the students by name. But no—these were undergraduates whose mentor happened to be (until the Kepler Mission came along) the most prolific planet hunter in human history. He knew everyone's name.

This was not the future Geoff Marcy would have imagined for himself when he was starting out in astronomy. In 1983, the same year Bill Borucki wrote his first paper on searching for planets, Marcy found himself sinking into a depression. He was more certain every day that he would never make it as an astronomer. You wouldn't have guessed it by looking at his

Geoff Marcy (C. Rose)

résumé: undergraduate degree in physics and astronomy from UCLA; Ph.D. from the University of California, Santa Cruz; postdoctoral fellowship at the Carnegie Observatories, in Pasadena, California—the same prestigious institution where Edwin Hubble discovered the expanding universe, and where countless other astronomical superstars have worked.

But that was the trouble. Marcy's sense of self-doubt had begun in graduate school, where some senior astronomers had criticized his dissertation on magnetic fields in stars. "I didn't have any confidence in my own ideas," he recalled during a conversation in the mid-1990s. "I was in a depression. I was convinced I was an imposter, that I didn't belong with all of these high-powered people." At Carnegie, which took only the very best of the best, they were even more high powered. They'd taken him as well, of course, but Marcy was sure they'd made a huge mistake. Everyone around him was brilliant, he could see that easily enough. But it was clear to him that he simply wasn't smart enough to be an astronomer.

One morning, Marcy hit bottom. Here's how I described it in my 1998 book, *Other Worlds*:

Marcy dragged himself out of bed and into the shower as usual, but instead of turning off the water when he was finished, he just stood there thinking. He knew he had to get himself out of what had become a perpetual depression. "I'm not Einstein," he thought. "I'm never going to be. So what am I going to do—beat myself up over it for the rest of my life?" He recalled how as a kid he had had posters of the planets plastered on the walls of his bedroom and had stayed up half the night for the pure joy of exploring the universe through his telescope and then he sat glued to the television to watch humans take the first steps on the Moon in July 1969. If he could somehow reconnect with the sense of wonder he had felt back then, he might be able to get excited about astronomy again. "I have to find something to work on," he told himself, "that addresses a question I care about at a gut level." It also had to be something difficult to do. There wouldn't be much satisfaction or self-respect in solving an easy problem.

All of this went through his mind while the hot water poured down his back, and while it seems a bit too romantically tragic to be true—and while Marcy does have a flair for the dramatic—I believe it. He had no inkling of what Bill Borucki was doing at just about the same time at the Ames Research Center, several hundred miles to the north. But Marcy decided to take on the identical challenge. He knew, as Borucki did, that the question of whether other worlds existed out among the stars had held the imaginations of philosophers and scientists for thousands of years. Reconnecting with his sense of wonder would not be a problem.

The search for distant worlds would also meet his other requirement: It would be extraordinarily difficult to do. Like Borucki, he knew about astrometry—measuring the back and forth wobbles an orbiting planet would impose on its host star. His instinct was the same as Borucki's: Those measurements would be so hard, and the technology required so complex and expensive, that it might be decades before it ever happened, if it happened at all. Unlike Borucki, however, Marcy didn't choose to look for the dimming of stars as planets transited in front of them. In fact, he told me nearly thirty years later, "Transits never even occurred to me. I never thought about them at all."

That's not especially surprising. Borucki was an expert in measuring light, so he decided to search for planets with a technique that required high-precision light measurements. Marcy's career had had mostly to do with breaking light apart to see what it could reveal about the star that was emitting it. The formal name of this process is spectroscopy, and it's been a staple of physics and astronomy since the 1800s. Its roots go back even further: Chinese scholars suggested as far back as the 1000s that rainbows were created when drops of water in the air split sunlight into a spectrum of colors. Persian and Arab astronomers came to the same idea independently, and in the 1200s the English natural philosopher Roger Bacon experimented with glasses of water and crystals that split light just as a rainbow does. Isaac Newton used prisms to split sunlight as well. William Herschel, who won worldwide fame for his discovery of the planet Uranus in 1781, did his own experiments with light and color that led to his discovery of infrared light in 1800—the first hint that light comes in colors the human eye can't perceive.

Physicists now understand that light comes *mostly* in colors we can't see. If you think of light as a piano keyboard, with each note representing a color, you can think of the visible spectrum as the octave right in the middle. Infrared is lower in pitch than visible light. Microwaves are lower still, and radio

waves are the deepest bass notes. At the other end of the keyboard, ultraviolet light is just a little too high in pitch for us to
see. Gamma rays are higher, and X-rays still higher—the shrill,
tinkly notes at the far right. (Like most analogies, this one isn't
perfect. There are vastly more colors of light, most of them
invisible to us, than there are notes on a piano).

The Sun shines in virtually all of these colors, at least to
some degree. We can see only the colors of the visible-light
rainbow because these are the colors that penetrate our atmosphere most easily. We evolved to take maximum advantage
of the kind of light that's most available. If we want to see at
night, when there's no sunlight, we can put on night-vision
goggles that are sensitive to infrared light. Living things, including plants and people, glow in the dark, but mostly in the
infrared. We've also invented technologies for sensing ultraviolet, microwaves, gamma rays, and all the other colors of light.
When astronomers began using these, in the 1920s, they began
finding all sorts of cosmic phenomena they'd never imagined
before—including, in 1965, the light left over from the Big
Bang, which is now detectable only as microwaves. Ultraviolet astronomy, X-ray astronomy, radio astronomy, and gamma
ray astronomy are now distinct, though obviously related,
branches of science.

In the early 1800s, William Hyde Wollaston, an Englishman,
noticed that fine, dark lines interrupted the artificial rainbows
he had created with prisms. A German chemist named Josef
von Fraunhofer independently discovered the same thing.
Within a few decades, chemists understood that the lines were
caused by chemical elements in the Sun's outer layers. The

elements absorbed very specific colors of light on their way from the Sun out into space. The elements are like filters. They catch *very* specific colors, so not just "blue" but "the exact color of blue represented by a particular wavelength of light." That's how specific we're talking about. A given element or compound doesn't just absorb light at a single wavelength, but at many different wavelengths at once. It's like a bar code, with a pattern unique to each substance. If light is shining through many different elements or compounds at the same time, the multiple overlapping barcodes have to be untangled before you can figure out what you're looking at.

Still, by comparing the lines they saw in the Sun with the ones they were able to create in labs on Earth, scientists were able to figure out what the Sun is made of. (It's made mostly of the same elements Earth is made of, although in very different proportions.) So that was useful. But spectral lines turned out to have another crucial property. They're exquisitely sensitive motion detectors. You can think of light as waves of electromagnetic energy. In the visible spectrum, the more tightly packed the waves are—which is to say, the shorter the distance between them, or the shorter the wavelength—the closer a color is to the violet end of the rainbow. If the waves are more loosely packed, they're closer to the red end. The same goes when you go beyond the visible part of the spectrum: Ultraviolet light is more tightly packed than violet; gamma rays are more tightly packed than ultraviolet, and X-rays even more. The piano analogy works here too: a note sounds higher pitched because sound waves—which are simply pressure waves in the air—are more tightly packed together. Looser packing makes for a lower note.

All of this is pretty straightforward when the thing you're looking at or listening to is sitting still. But now imagine something that's making a lot of noise while it's coming toward you—a train speeding toward you with its horn blaring, for example, as you stand close to the track. As the train comes toward you, its motion squeezes the sound waves together, so the pitch sounds higher to you than it really is. When the train passes and starts moving away the squeezing stops abruptly, and suddenly the sound waves are being stretched instead. The pitch drops instantly (in old movies, train whistles are doing this all the time; these days, you mostly hear the effect with police and ambulance sirens).

This change in pitch works for things that are moving toward or away from you; if they're moving from side to side, there's no squeezing and no stretching. So you need to be right next to the railroad track to hear the switch from high pitch to low.

Exactly the same thing happens with light. If a shining object is moving toward you, its light looks slightly higher-pitched than it really is. That is to say, it looks bluer. If it's moving away, it looks redder. If a star happens to be moving toward you, it isn't just the light, but also the dark lines that interrupt it, that shift in the blue direction. This trick of light is what let Edwin Hubble discover the expanding universe back in the 1920s. When he broke apart the light from galaxies beyond the Milky Way, he thought he would see shifts in the locations of dark spectral absorption lines to show that some of the galaxies were moving away from us and some moving toward us. Instead, the lines showed that they were all speeding away. This meant either that the Milky Way was somehow very repulsive, or, more reasonably, that the whole universe was expanding, with every galaxy speeding away from every other galaxy.

During graduate school and on into his postdoctoral fellowship, Geoff Marcy had gotten very good at finding and measuring spectral lines. So he went with his strength and decided to look for planets this way. With astrometry, you have to be looking down on an alien solar system from above to see the star moving from side to side ("down" and "above" don't really have any meaning in space—you could just as easily say "up from below," and you'd be equally inaccurate, but it's such an instinctive way to describe it that astronomers talk this way anyway, and nonscientists understand instantly what they're talking about).

Marcy wanted to look for that same motion, but from an edge-on perspective. He wanted to catch planets tugging their stars toward Earth (just the tiniest bit), then away. Since the red-shifting and blue-shifting of spectral lines betrays that sort of motion, that's what he proposed to look for.

The only problem with this idea, Marcy soon learned, was that it was nearly impossible. When cosmologists use shifting spectral lines to measure the speed of galaxies racing apart as the universe expands, they're looking at objects moving at many thousands of miles per hour. They take a spectrum from a galaxy's collective starlight, lay it next to a reference spectrum from a motionless object—the Sun, for example, or a laboratory reference lamp that generates an artificial spectrum—and see how far a given line has shifted. There's some imprecision in the process, but if they're off by a couple of thousand mph or so, that's plenty accurate enough. They don't have any need to improve their precision.

But the back-and-forth motion Jupiter causes in the Sun is a piddling 28 mph or so. The greatest expert in measuring cosmological redshifts would fail utterly to detect it. So Marcy tried tightening up the procedure in every way he could think of to make it more precise. He succeeded up to a point: He managed to get his accuracy down to about 450 mph. But since he was trying to find distant Jupiters, this wasn't nearly good enough. No matter how careful he was, the act of moving the telescope from star to reference lamp changed his measurement system enough to make the measurement unreliable. Imagine you wanted to measure the length of two different objects—two bricks, say—with high precision. You'd be smart not to use two different rulers, since one of them might be just a little bit off. But if you wanted to be *really* precise, it's a problem even if you use just one ruler—moving the ruler from one brick to the other could change things. The second brick could be in a slightly warmer place, so the ruler might expand just an infinitesimal amount. Or you might hold the ruler in a slightly different way, so it would sag under gravity differently, distorting its shape. These are absurdly small changes, but if you really needed absolute accuracy, they could make a difference. The best way to make absolutely sure you're measuring things exactly the same way is to measure them at exactly the same time.

Marcy decided to do just that. He'd measure a star's spectrum and a reference spectrum all at once, so he didn't have to move anything. He might have figured out a way to do it, but it turned out that he didn't have to. A Canadian postdoc named Bruce Campbell, at the University of British Columbia, had come up with a solution a half decade or so earlier. Working with a colleague named Gordon Walker, Campbell had realized that you could take a gas whose spectral absorption lines were thoroughly understood, put it in a glass container, and let starlight pass through the gas on its way to the spectrometer. The reference spectrum and the real spectrum would be measured at exactly the same time with exactly the same instrument.

It worked. Campbell and Walker were able to measure the wobbling of stars to an accuracy of plus or minus 30 mph or so. That still wasn't precise enough to find an alien Jupiter, however, since the measurement error was as big as the signal you'd be looking for. Beyond that, Campbell and Walker had settled on hydrogen fluoride gas for their reference—well understood, but also corrosive, explosive, and horribly toxic. Marcy needed a better gas, and he also needed a collaborator for this project, which was growing increasingly complicated.

By now, Marcy was on the faculty at San Francisco State University. After asking around a bit, he learned of a recent San Francisco State graduate who had joint degrees in physics and chemistry, and who also had a strong interest in astron-

omy. Paul Butler was still at the university, working on a master's in physics and looking for a research topic. When Marcy approached him, Butler was intrigued. Like Marcy, Butler was drawn to research with long odds but a potentially huge payoff. And he loved the challenge of trying to make measurements more precise than anyone had ever been able to pull off.

The only downside was that Paul Butler had a somewhat rough personality. He divided the world into good guys and bad guys, and the bad guys included some of the world's most eminent scientists. These would eventually include, once he signed on with Geoff Marcy, many of their professional colleagues. He thought nothing of describing senior astronomers at places like Caltech and Cornell and especially Harvard as evil, or cowardly, or even mentally ill. He would say these things openly, and later on, when Marcy and Butler finally began making discoveries and getting some public recognition, he would sometimes even say them to reporters.

Nevertheless, Butler was very good at his job. He spent more than a year hanging out with chemists, trying out one element or compound after another, looking for the ideal reference gas to use for finding planets. Ultimately, he settled on iodine. It was not only safe, but it also had an enormous number of spectral absorption lines that spanned the visible spectrum all the way from red to violet. That would give each measurement plenty of cross-checks. The lines created by hydrogen fluoride, by contrast, were not only fewer in number, but they also bunched up in a small part of the visible-light spectrum.

So Marcy and Butler built what they called an "iodine cell" to attach to the Hamilton Spectrograph at Lick Observatory near San Jose, and began using a small telescope to take data on relatively bright, nearby Sun-like stars, looking for wobbles. They didn't have the software yet to analyze their observations; the spectrum of iodine was so horribly messy that they couldn't disentangle its spectrum on their images from the spectra of the stars. But Butler was also a talented software writer, so while they continued to take unreadable measurements, he worked on code that might someday make sense of them.

In the end, it took him six years. "It's my Rembrandt," Butler told me in 1996. "It's as close to great art as I'll ever get." Even then, however, Marcy and Butler could get to a precision of only twelve meters per second—still not good enough to find an alien Jupiter. The Hamilton Spectrograph, built by Steven Vogt, Geoff Marcy's thesis adviser in grad school at Santa Cruz, was now the limiting factor. Vogt had to upgrade the device, and then Butler had to rewrite his software to account for the upgrade. Finally, Geoff Marcy, Paul Butler, and Steven Vogt, their new collaborator, were able to measure the wobbles of stars to within an astonishing and unprecedented three meters per second. They could find a planet like Jupiter.

Chapter 3

HOT JUPITERS:
WHO ORDERED THOSE?

A	FTER GEOFF MARCY and Paul Butler had all the
	kinks worked out in their hardware and software, the
only thing standing in their way was time. It takes Jupiter
eleven years to orbit the Sun just once. If you were an alien
astronomer looking toward our solar system using an iodine
cell and the Hamilton Spectrograph, it would take you eleven
years to watch the Sun move toward you, then away, then
back in a single orbital cycle. And if you were a really careful
alien astronomer, who didn't want to risk the embarrassment
of making a discovery that turned out to be wrong, you'd want
to see not just one, but at least two or three cycles to convince
yourself you really were seeing a planet and not, say, some
weird pulsation of the star itself.

Geoff Marcy knew very well that astronomers had fooled
themselves about planets before. The best-known example was
the "discovery" of planets around a nearby star known as
Barnard's Star, by Swarthmore College astronomer Peter van
de Kamp in the 1960s. What van de Kamp thought was a

side-to-side motion in the star, caused by a planet, was actually a change in his telescope—a minuscule repositioning of a lens when the telescope was refurbished. The slight change in focal length made Barnard's Star appear to move, and van de Kamp interpreted the motion as the tug of a planet. When the mistake was discovered in the 1970s, van de Kamp compounded the problem by being slow to acknowledge it (in fact, it's not clear that he ever did).

To be sure they could claim a planet discovery with confidence, Marcy, Butler, and Vogt would have to wait a long time to confirm that any wobble they spotted matched the signature of a planet. They'd also have to convince other astronomers that their complicated apparatus and their complicated software—which may have been Butler's Rembrandt, but which was so complicated that it would take years to analyze observations of a single star—really was capable of doing what they claimed. Some of their colleagues at other universities literally laughed at them when they heard a description of the research. Many of the laughers ended up on Butler's growing enemies list.

Near the top of that list was an astronomer named David Latham. Latham has been at Harvard since the late 1960s; he got his Ph.D. there in 1970. When I arrived as a freshman in the fall of 1971, he was the deputy instructor of a hugely popular undergraduate course called The Astronomical Perspective, taught by the astronomer and historian of science Owen Gingerich. I remember looking down from the back row of a steeply angled lecture hall and seeing Latham, a young, skinny, nondescript guy, standing off to the side as Gingerich lectured.

Forty years later, I sat in his office at the Harvard-Smithsonian Center for Astrophysics—the CfA, if you want to sound like an insider—listening to an older guy, less skinny, with grayer hair, wearing a jacket and tie (he hadn't worn them as a teaching assistant in 1971, and most astronomers don't wear them now unless they're getting a major award). He was still teaching the course I'd taken so many years earlier, but it now had a different name, and he was now the lead instructor, since Gingerich had retired. Latham was still pretty nondescript, but now it was in an avuncular way—low key, companionable, easy to get along with. You wouldn't immediately guess, nor would you have guessed in 1971, that he was and is a competitive motocross racer (a photo of Latham kicking up dust on his dirt bike appears on his Harvard homepage) and a hockey player. It's less surprising to learn that he's a wine connoisseur.

About midway between his appearance at the front of that lecture hall in 1971 and our conversation in 2011, Dave Latham had become a planet hunter too. Originally, he was largely interested in cosmology, the study of the origin and evolution of the universe. The Big Bang had become the dominant theory of the universe in the mid-1960s, after lingering around the edges of science for many decades, but there were all sorts of profound unanswered questions remaining. When did the universe begin? What happened during the first moments after the Big Bang? What is the universe made of? What will its ultimate fate be?

During the 1970s, it had also became apparent that the universe was pervaded by some sort of mysterious, invisible substance, at first known as the "missing mass" and later called "dark matter" (that mystery still hasn't been solved). The dark matter would have had a powerful influence on how the cosmos evolved, so astronomers wanted to understand how galaxies were spread throughout the universe. Were they sprinkled evenly or were they assembled into some sort of pattern that hinted at how much dark matter there was and how it was distributed? (We now know they're in patterns that resemble Swiss cheese, which helps to rule out some models of cosmic evolution.)

Latham was involved in some of the early surveys that tried to understand galaxy distribution. The galaxy surveys, operated out of the Smithsonian's telescopes on Mt. Hopkins, near Tucson, Arizona, used spectrographs. But galaxies are too faint to see when the Moon is bright in the sky. Latham hated to see the instruments sit idle, so when the Moon sidelined the galaxy surveys, he began looking at stars in the Milky Way to see if they wobbled. He wasn't interested in planets at this point, but in binary stars—pairs of stars that orbit each other. These are actually more common than single stars in the Milky Way. The Sun is unusual in wandering through the galaxy alone. "We were looking to see how frequent binaries were," said Latham, "and what their characteristics were. By the eighties we were mass-producing radial velocities [the technical term for motion toward and away from the observer]. We had our errors down to maybe five hundred meters per second," which is about a thousand miles per hour.

Then, in 1984, an Israeli astronomer named Tsevi Mazeh contacted Latham. Mazeh was interested in radial-velocity measurements too. He had just been out in Santa Cruz, California, consulting with Steve Vogt. Geoff Marcy had barely begun thinking about planets at this point. Mazeh was on his way back to Tel Aviv, but he stopped in Massachusetts en route. "He had this idea," said Latham, "to use my radial-velocity instrument to search for giant planets." Latham patiently explained to Mazeh why this wouldn't work, that even a planet like Jupiter would pull on its star too weakly to show up, and that it would take more than a decade to see a single orbit in any case. "No," said Mazeh, "I mean giant planets with very short periods."

You could certainly find those: A short orbital period means you can watch several orbits go by without waiting forever. The planet Mercury orbits the Sun once every eighty-eight days. If a giant planet hugged its star the way Mercury does the Sun, you could see four full orbits in just under a year. Beyond that, a planet's gravitational effect on its star is more powerful the closer it orbits—it has more leverage—so the observations don't have to be nearly as delicate. That was all true, acknowledged Latham, but giant planets don't orbit that close to stars. They certainly don't in our solar system, and most planetary theorists were confident they couldn't exist at all.

"Maybe," said Mazeh, "the theorists are wrong."

"He seemed like a nice fellow," recalled Latham, "so I agreed to work with him." To boost their chances further, Mazeh suggested they look at M-dwarfs, red stars that are at most half as massive as the Sun. M-dwarfs make up about 70 percent of the stars in the Milky Way, but they're too dim to see with the naked eye. The fact that they're such lightweights means that a giant planet orbiting an M-dwarf would make the star wobble even more, making it that much easier to detect. "We also decided to look at some Sun-like stars," said Latham. "On the night of March 31, 1988," he continued, "I was working up the observations on one of them, a star called HD 114762."

At this point, readers should be warned that the naming conventions for stars are complicated enough to make your head hurt. The HD in the name of the star Dave Latham was looking at signals that it appears in the Henry Draper catalog, a list of more than 350,000 of the brightest stars visible from Earth, classified according to features in their spectra. (Draper, a medical doctor and amateur astronomer, took some of the earliest photographs of a star's spectrum. The catalog named in his honor was assembled by Harvard astronomer Edward Pickering, using funds donated by Draper's widow in the 1880s.)

But that's just one of many star catalogs. Another is the Gliese catalog of the closest, rather than the brightest, stars. The German astronomer Wilhelm Gliese put it together in 1957. "Gliese names," Geoff Marcy told me, "are kind of nice, because they tell you right away it's a nearby star even if you don't know anything else. But the HD catalog is more widely used." And then there are the Hipparcos and the Tycho catalogs, gathered by the European Space Agency's Hipparcos satellite in the 1990s (the satellite's mission was to map the positions of millions of stars; the Hipparcos catalog has very high precision, the Tycho catalog is based on somewhat less careful measurements with the same satellite).

It doesn't end there. Many stars are also identified by the constellation they're part of. Alpha Centauri is the brightest star in the southern constellation Centaurus (*alpha* is the first letter in the Greek alphabet). If a constellation has more than twenty-two stars, you run out of Greek letters, so you go to numbers. The constellation Pegasus has a star known as Alpha Pegasi, and it also has one called 1 Pegasi. Then there are stars with given names—Vega, Sirius, Aldebaran, Betelgeuse, Polaris, Arcturus, and more. These stars are so bright and prominent in the night sky that ancient Greek and Arab sky watchers thought of them as old friends, with distinct personalities. Finally, when astronomers are conducting an organized search, they'll sometimes create their own catalogs—the Kepler catalog, the Wide Angle Search for Planets (WASP) catalog, the Hungarian Automated Telescope (HAT) catalog, and so on. According to one astronomer I spoke with, this is partly to make sure the credit for a planet's discovery goes prominently to the search team.

The one catalog astronomers don't pay attention to is the one compiled by the International Star Registry. This is the company that lets you "name a star after someone." The ads say, quite truthfully, that the names will be recorded in book form in the Library of Congress. But that's true for any book, whether it's an erudite volume of history or a trashy romance novel. The company makes clear on its website that astronomers don't consult the book.

Even when you're talking about stars identified by their numbers in a single catalog—in this case, the Henry Draper catalog—the names begin to swim before your eyes. Unless you're a planet hunter, that is. Then they're as distinctive as the names of your children. "Exactly!" said Debra Fischer, a

Yale astronomer, when I proposed this analogy. "You don't know HD 209458? These names are burned into my memory. Someday I will have Alzheimer's, but I will remember these stars." For the record, Fischer tends to rely on the Hipparcos catalog, because it has more than two million stars; because it notes their positions and distance from Earth with exquisite accuracy; and because all of the information is online, making it easy to access.

So Latham was working on HD 114762, a star in the Henry Draper catalog. His calculations suggested a wobble with a period of about eighty-four days, about the same as Mercury's. The magnitude of the wobble suggested an object with a mass about eleven times that of Jupiter—or rather, that was the *minimum* possible mass. This was a crucial point. Since the object pulling on the star was itself invisible, Latham couldn't be certain its orbit was truly edge-on. If it was, the mass was eleven times that of Jupiter. If the orbit had been at right angles to Latham's telescope, you wouldn't see any motion at all toward or away from the telescope; it would all be side to side. But if the orbital plane was tilted somewhere in between those extremes, you'd see something between the full effect and zero. Any motion toward or away from the telescope would reflect *part* of the tugging. A much bigger object, pulling at an angle, could mimic an eleven-Jupiter-mass body seen directly edge-on.

Latham sent an e-mail to Mazeh, with a copy to Michel Mayor, a Swiss astronomer who was also looking for wobbling stars. Mayor wasn't looking for planets either; he was looking for brown dwarfs, objects bigger than planets but smaller than stars. A brown dwarf could be as much as eighty times as massive as Jupiter before it would start fusing hydrogen into helium in its core—the same reaction that powers an H-bomb, and the one that makes stars shine. At this point, brown dwarfs were purely theoretical (they've since been shown to exist), but looking for them had made Mayor an expert on radial-velocity measurements too. The e-mail said, in part, "This is interesting—the minimum mass is well under the stellar limit. It could even be a giant planet." By the time Latham hit "send," the clock had ticked past midnight. It was now April 1, and he was a little bit worried that Mazeh and Mayor might think it was an April Fool's joke.

But Mayor went out and did his own measurements, and got the same results. Mayor also determined that whatever this object was, its orbit was eccentric—it was somewhat oval rather than nearly circular. That ruled out a planet, because, as Latham said, "everyone knew that giant planets had to have circular orbits." Everyone also knew you couldn't have planets bigger than about twice the mass of Jupiter. And with a "year" just eighty-four days long, well, said Latham, "that was just a killer. Three strikes, you're out! Tsevi and I had a bet—we still do, in fact—I said it was a small star, he insisted it was a big planet." When Latham, Mayor, Mazeh, and two others reported the discovery in *Nature* on May 4, 1989, they wrote: "The companion is probably a brown dwarf, and may even be a giant planet." Latham told me Mayor "wasn't too happy with that wording." He was on Mazeh's side.

In any case, Latham was too busy with other things to keep looking for objects like this one, but Mayor, he says, "picked up [the project] and ran with it." Like Marcy and Butler, he and a French instrument builder named Andre Baran began beating down the errors in their own spectrograph. Marcy and Butler had chosen to do it with iodine cells and horrifically complex software. Mayor and Baran chose instead to make their spectrograph as utterly stable as they could. They used a reference spectrum from a lamp outside the telescope, but they piped it into the spectrograph with fiber-optic cables in such a way that it was as undistorted as it could possibly be.

When they'd done everything they could think of, Mayor told me at a conference on the Isle of Capri in 1996, they had beaten down their errors to thirteen meters per second, or about 30 mph. They couldn't find a Jupiter like the one in our solar system, but then, they weren't looking for one. Mayor still cared mostly about brown dwarfs, and the spectrograph, far more sensitive than Dave Latham's, could detect them easily. In early 1994, Mayor and a graduate student named Didier Queloz began taking measurements of wobbly stars. By now, Marcy and Butler had been at it for half a dozen years.

Mayor and Queloz had put more than a hundred stars on their observing list to maximize their chances of finding something. Brown dwarfs might be relatively rare, after all, and their orbits would have to be nearly edge-on for the astronomers to make a strong detection. Not all of them would be, of course. So the European astronomers began methodically ticking through their list. Within a few months, they noticed something very odd. A star named 51 Pegasi (the fifty-first brightest star in the constellation Pegasus) seemed to be wobbling, but in an impossible way. It was moving back and forth, not with a Jupiter-like rhythm of 11 years, not with an Earth-like rhythm of 365 days, not even with a cadence of 84 days, like HD 114762. This star was moving toward Mayor's telescope and dancing and advancing again *once every four days*. If this motion were truly caused by an orbiting body, it was hugging its star an absurd ten times closer than Mercury hugs the Sun.

Not only that, but based on how hard it was yanking on the star, this was no brown dwarf: It was only half as massive as Jupiter. Mayor's spectrograph wasn't sensitive enough to find a Jupiter in a Jupiter-like orbit, but something this close in had a huge amount of leverage on the star. It looked just the way a giant planet should look—if a planet could exist in this location. Theorists said it couldn't. But as Tsevi Mazeh had said ten years earlier, "maybe the theorists are wrong."

In fact, not all theorists had said such a close-in planet couldn't exist. Douglas Lin, of the University of California, Santa Cruz, had proposed back in 1992 that giant planets might migrate inward from where they originally formed. He figured they'd just spiral all the way into the star and be destroyed—but for a while, they could take up an orbit like the one Mayor was describing. And forty years before that, the legendary theorist Otto Struve had written a paper for the October 1952 issue of a journal called *The Observatory* titled "Proposal for a Project of High-precision Stellar Radial Velocity Work," in which he foreshadowed not only Mayor's and Marcy's and Dave Latham's work, but Bill Borucki's as well. In part, Struve wrote:

We know that *stellar* companions can exist at very small
distances. It is not unreasonable that a planet might exist at
a distance of 1/50 astronomical unit, or about 3,000,000 km.
Its period around a star of solar mass would then be about
1 day. If the mass of this planet were equal to that of
Jupiter . . . [it] might be just detectable . . . There would,
of course, also be eclipses . . . This, too, should be ascer-
tainable by modern photoelectric methods.

Struve's idea had been largely forgotten, however, and Lin's
work was considered highly speculative. If an observer is
wrong as much as 10 percent of the time, goes the astronomical rule of thumb, he or she is a pretty careless observer. But if
a theorist is wrong as little as 10 percent of the time, he or she
isn't taking enough creative risks. Lin was, and remains, a very
good, creative theorist, so his colleagues took his predictions
with a grain of salt.

Mayor and Queloz returned to the telescope, trying to
make absolutely certain that they weren't somehow kidding
themselves. Maybe the star was pulsing, its outer atmosphere
bulging toward and then away from Earth in a four-day
rhythm. Maybe the star wasn't perfectly spherical, and they
were seeing the bulgy part moving toward them and then
away. "The first principle [of science]," the physicist Richard
Feynman said in a commencement talk at Caltech in 1974, "is
that you must not fool yourself—and you are the easiest person to fool." Peter van de Kamp had fooled himself with his
"discovery" of a planet orbiting Barnard's Star (he fooled
others as well; Otto Struve's 1952 paper refers to "results

announced . . . by P. Van de Kamp"). Mayor and Queloz, like Marcy and Butler half a world away, were determined to avoid destroying their reputations. But hard as they tried to make the impossible conclusion go away, it refused.

In the end, Mayor told me on Capri, "It's a difficult thing to decide you've done all you can, that you're ready to leave your office and go public." They submitted a paper to the journal *Nature* claiming the discovery of a planet-like object they called 51 Pegasi b (the *b* meaning that it is a secondary object orbiting the star 51 Pegasi). Before the paper could be published, he spoke about the discovery at a conference in Florence, Italy, in October 1995. According to *Nature*'s strict rules, he was allowed to do this, but he wasn't allowed to discuss the findings with reporters until the paper was actually published. If he did so prematurely, *Nature* wouldn't publish it after all — and *Nature* was prestigious enough that Mayor didn't want to flout the rules. There were reporters at the Florence conference who begged him for interviews. He politely refused, so they went ahead and announced the discovery without quoting the man who had made it.

In California, meanwhile, Geoff Marcy and Paul Butler began hearing about Mayor's find, first from colleagues who had been at the meeting, and then from reporters who were desperate to find an expert they could talk to. It seemed obvious to Marcy that Mayor had made a mistake. The sort of planet he was describing couldn't possibly exist. The theorists said so. Besides, Marcy couldn't possibly be scooped: He and Butler had been working tirelessly to find planets for six years now. How could someone else just stumble onto the discovery?

Still, he wasn't going to say Mayor was wrong without being absolutely sure. Astonishing things often turn out to be
false—but not always. In 1989, for example, two chemists from
the University of Utah claimed they'd discovered "cold fusion," an inexpensive source of potentially limitless clean energy. I called Rob Goldston, the director of the Princeton
Plasma Physics Laboratory, who was struggling to create fusion in a multibillion-dollar installation owned by the Department of Energy. "I don't know the details of the experiment,"
he told me, "so I can't make any definitive statement." "But," he
continued, making it clear as diplomatically as possible how
he really felt about the claim, "if it's true, it means that everything we've learned about nuclear physics over the past fifty
years is false." In other words, Goldston was almost certain
the chemists were wrong, but knew that an absolute statement
might come back to bite him.

Marcy and Butler were convinced Mayor must be wrong.
But maybe everything they knew about planets was false. Either way, they were all set up to find out for themselves. If
Mayor's instruments could see this "planet," theirs could too.
To find a planet in a four-day orbit, you have to look at least
once a day; Marcy and Butler had never bothered to look at
any star more often than once every few months, because the
wobbles they were looking for would play out over years. So
they went up to Lick Observatory, where they'd already been
approved for four nights on the 120-inch reflecting telescope.
As the data streamed in from 51 Pegasi, they would immediately funnel it into their computers for processing, taking more
data all the while.

After a couple of days, they knew Mayor was right. They'd been scooped by someone with a less sensitive instrument. Arguably, they'd been scooped by Dave Latham back in 1989 as well, but Latham's discovery had never been accepted as a planet, even by Latham himself. Ultimately, all the strikes against Latham's object would be eliminated as astronomers began to understand how strange planets really could be. In hindsight, Geoff Marcy now gives Dave Latham credit for the very first planet orbiting a Sun-like star.

Michel Mayor couldn't talk to reporters, but Marcy and Butler were under no such obligation, since they didn't have a paper about to come out. Reporters couldn't talk to Mayor, so they descended on the Californians. And while Mayor had made the discovery, Marcy and Butler *could* have made it. If luck had been on their side, they inevitably would have. So it was a sort of poetic justice that Marcy and Butler were the ones who ended up being lavished with public recognition. Geoff Marcy also realized that if they'd known how comparatively easy the universe would make it to find planets, they might not have worked so hard to make the Hamilton Spectrograph so precise. Their ignorance had put them behind, but it had given them an edge on future discoveries.

In fact, the discoveries had already been made, even if Marcy and Butler didn't know it. The astronomers had been searching for wobbles for many months now, returning over and over to a list of 120 stars. But they hadn't bothered analyzing the data, since they thought no star would show a perceptible wobble over so short a time. The information was just sitting in storage on magnetic tapes. Now that they knew such a thing was possible, however, they realized the signals of planets might be on those unread tapes. Butler was "insane," he later told me, to find out what was on them. He got his hands on some relatively fast computers and began feverishly processing data around the clock.

Within less than two months, Butler managed to tease out the signal of a planet orbiting the star 47 Ursae Majoris, in the Big Dipper. Then he found another, orbiting 70 Virginis, in the constellation Virgo. Neither of them was as crazy as 51 Peg b, as astronomers were now nicknaming it, but they were still closer to their stars than Jupiter is to the Sun. They were still a little crazy. 70 Virginis b, in particular, had an unusually eccentric, egg-shaped orbit. That was one of the strikes Dave Latham had listed against HD 114762.

Nevertheless, Geoff Marcy arranged to give a talk on their new planets at the winter meeting of the American Astronomical Society, which was coming up in a few weeks. Unlike Michel Mayor a few months earlier, he and Paul Butler didn't have a paper ready for publication, so they weren't bound to avoid reporters. They gave a heads-up to the society's press officer, an astronomer named Steve Maran, who immediately scheduled a press conference. He wouldn't say in advance what it would be about, although he told me privately, and undoubtedly told other reporters as well, that I'd be crazy to miss it.

Despite the veil of secrecy, however, word got out that Marcy would have something important to say, and when he finally did, he spoke with a theatrical flair that I would later recognize to be his trademark. "After the discovery of 51

Pegasi b," he said, "everyone wondered if it was a one-in-a-million observation. The answer is . . . no. Planets aren't rare after all." He went on to describe 47 UMa b and 70 Vir b. He also pointed out that the latter orbited in the habitable zone of its star, the region where water could exist in liquid form, the necessary ingredient for life. And while 70 Vir b itself was too big to support living beings (it's about six times as massive as Jupiter), it might have habitable moons. This was pure speculation, but it was a bold enough statement to get the discovery on the cover of *Time* and into headlines and news broadcasts around the world.

What all those viewers and readers didn't realize was that Marcy and Butler's announcement marked an enormous change in the way scientists would think about extraterrestrial life from that moment forward. For about two thousand years, philosophers and scientists had actively debated the question of whether life exists beyond the Earth. From the Renaissance on, it was widely believed that the answer was yes. But since the early 1900s, when astronomer Percival Lowell convinced himself that he could see canals and other evidence of life on Mars, thinking about and looking for life on other planets had been considered something of a fringe idea in science. The UFO craze that started in the 1950s didn't help.

In principle, scientists thought it was plausible that life existed elsewhere in the universe, and a few even tried looking for it—Frank Drake, who began the formal Search for Extraterrestrial Intelligence in 1961, for example, and a few NASA scientists who designed biology-sensing experiments for the Viking Mars landers in the 1970s. But no one even knew for sure that there were any planets beyond the solar system for life to exist on, and it was clear to most astronomers that planets were just too difficult to find with existing technology.

Suddenly, because Marcy had ignored the conventional wisdom, and because Mayor had gotten lucky, it was clear that this simply wasn't true. It's something like what happened when the British athlete Roger Bannister ran the world's first sub-four-minute mile in 1954. Until he did it, many people thought it was impossible. Once Bannister showed otherwise, plenty of runners found that they could break four minutes as well. Those first three planet discoveries, considered impossible by many scientists, forced the entire field of astronomy to shift its perspective. Men and women who had gone to graduate school planning to study other topics changed direction and decided to look for planets instead. Senior astronomers who had spent their careers thinking about the Big Bang or the formation of galaxies did the same.

This influx of brainpower and of funding from NASA and other agencies led in turn to ingenious new ways to search, which turned up dozens, then scores, then hundreds of new planets, in such a bewildering array of sizes, shapes, and orbits that astronomers are still arguing, a decade and a half later, about how solar systems form and evolve. Even so, by the time Bill Borucki's lunch was delayed by an AP reporter's nagging questions in early 2011, no one had yet found a true twin of Earth—the likeliest place, given the admittedly little we know about extraterrestrial biology, where life might actually be found.

It wouldn't be long, though. And while many of the astronomers working on the Kepler project had been inspired to go into planet-hunting by Geoff Marcy's and Michel Mayor's extraordinary discoveries in the 1990s—and even though Marcy himself would sign on as a project scientist with the Kepler Mission before the spacecraft launched—neither Marcy nor Mayor might end up playing a direct role in that discovery.

Chapter 4

AN ANCIENT QUESTION

WHEN BILL BORUCKI and Geoff Marcy set out in-dependently to find worlds orbiting other stars, other astronomers were dubious only because it seemed obvious that astronomical technology wasn't yet sophisticated enough to find them. They had no doubt that such worlds existed. The almost universal attitude among astronomers, and among most other scientists who stopped to think about the question, was, "How could there *not* be planets around other stars?" The argument was both numerical and Copernican. The Copernican part refers to Nicolaus Copernicus, who showed in the 1500s that the Earth was not, as European philosophers believed, at the center of the universe. That discovery has now been generalized to argue that the Earth, and the Sun, and the solar system, and even the Milky Way galaxy, aren't special in any way that matters to the universe. If planets exist in our solar system, the Copernican principle suggests they probably exist around at least some other stars as well.

The numerical part comes from the fact that the Milky Way is made up of at least one hundred billion stars, and possibly as many as three hundred billion. Most of the stars are smaller than the Sun—they're dim, red objects known as M-dwarfs—but even if you set those aside, tens of billions of stars very much like the Sun remain. It's possible that some of these are wandering through space unaccompanied by planets. It's impossible, or at least it seems very, very improbable, that all of them are.

When the question about planets arose thousands of years ago in ancient Greece, its meaning was a bit different. Nobody understood back then that the Sun was just another star, much closer to us than the rest. Nobody knew that the Milky Way, which split the night sky, was made up of billions of stars so far away that their light merged into a glowing band across the sky. For the Greeks, *world* and *cosmos* were interchangeable: The world included the Earth plus the sky and everything in it. So when the philosopher Epicurus wrote in the fourth century B.C. that "there are infinite worlds both like and unlike this world of ours . . . There nowhere exists an obstacle to the infinite number of worlds," he was talking about what we would now call parallel universes, places forever inaccessible to us that would consist of a parallel Earth whose skies were dotted with glittering stars.

Epicurus and his rough contemporaries Democritus and Leucippus were known as "atomists," because they believed the world and everything in it were made of tiny, invisible, and indivisible units they called atoms. These atoms, they argued, gathered together in different ways to form the stars, the Moon, the Sun, the planets, and the Earth, along with everything on it. When physical objects changed—the wood in a fire turning into ashes, flames, and smoke; the wood in a tree growing larger and bulkier every year; the wood in a downed tree slowly rotting away—it was simply a matter of the atoms rearranging themselves into another form.

It all sounds impossibly modern, until you realize that atomism was a purely philosophical idea. The atomists had no evidence of any kind to support their assertions. They did no experiments that could test the hypothesis, so there was no way to refute it if it was wrong. That made atomism unscientific by definition, and it remained so until the early 1800s, when the English chemist John Dalton revived the idea to explain the nature of the chemical elements. Unlike the Greeks, Dalton and the scientists who came afterward did all sorts of experiments that could have demolished the atomic theory. The theory survived, obviously, although it wasn't fully confirmed until 1908 by the French chemist Jean Baptiste Perrin.

With infinite numbers of atoms swirling around an infinite void, the atomists reasoned that it was inevitable for them to form into infinite numbers of worlds. Writing six hundred years later, the philosopher Diogenes Laertius paraphrased the atomist Leucippus as believing that "the worlds come into being as follows: many bodies of all sorts and shapes move by abscission from the infinite into a great void; they come together there and produce a single whirl, in which, colliding with one another and revolving in all manner of ways, they begin to separate like to like."

Again, it seems remarkably sophisticated. Leucippus might have been describing the modern theory of planet formation, in which a cloud of interstellar gas and dust collapses under gravity to form a dense knot of matter—a future star—surrounded by a swirling disk of material that will turn into planets. But again, this wasn't a scientific idea but rather a philosophical one. And Leucippus's worlds, like those of Epicurus, were entire, self-contained universes, entirely separate from one another and constructed in different ways. (In another seemingly uncanny foreshadowing of modern science, this concept predates by more than two millennia the multiple-universe theories that have arisen independently from string theory, quantum theory, and inflationary cosmology.)

Given the resonance of their ideas with those of twenty-first-century physics and cosmology, you might expect the names of Epicurus, Democritus, and Leucippus to be more familiar than they are. Instead, we're familiar with Aristotle, who lived at about the same time. That's because his theory of the universe won out over theirs, and in Aristotle's theory, only a single world was possible. In Aristotle's universe, everything was made of just four elements, not an infinite number of atoms. The elements didn't swirl around, turning from one thing into another: They moved, inexorably, to their natural resting places and (mostly) stayed there. The heaviest of the four was earth—perfectly sensible, since the ground always lies under our feet and objects made entirely or partly of earth (rocks, trees, people) tend to fall to the ground unless something prevents them from doing so.

The lightest element in Aristotle's cosmology was fire. That makes sense as well: Things made of fire (lightning, shooting stars) are high in the sky, and the flames of an actual fire try to leap upward, to their natural place. In between earth and fire were, from the bottom up, water (it sits on top of the ground in the form of puddles, lakes, rivers, and oceans, and it falls to the ground as rain) and air (bubbles rise in water, and the air quite clearly sits above lakes and oceans but below the Sun).

If Aristotle's cosmology was right, the atomists' notion of other worlds had to be wrong. The natural home for the element earth was under our feet; if there happened to be any substantial amount of it up in the heavens, it would long since have returned home. To imagine other worlds, you'd have to imagine multiple natural homes, which made no sense. Even if those homes existed, for the sake of argument, how would the air know where to go? If it migrated up to its natural place above our Earth, it would be moving down toward an unnatural location with respect to another Earth. It was—assuming Aristotle was right—crazy.

Aristotle's theories won out not only because they were intuitively persuasive, but also because they described an orderly universe, not a swirling chaos of invisible atoms randomly coming together. They were also comprehensive, not only describing the makeup of the world but also explaining the motions of the Sun, Moon, and planets (they were embedded in nested crystalline spheres that rotated majestically around the Earth at different rates). It was a neat, self-consistent cosmology, based on real things people could see. This turns out to be another foreshadowing of the modern controversy over string theory. It's the hottest idea in modern physics: The building blocks of matter aren't subatomic particles, but rather vibrating loops of "string," far smaller than the particles we know, which live in an eleven-dimensional space that's mostly invisible to us. It's a mathematically powerful theory, but there's no clear way to test it, aside from building a particle accelerator the size of the Milky Way. Some physicists argue that this makes it pure philosophy, not science.

In any case, Aristotle's cosmology, along with the rest of his science, won over enough of his colleagues and disciples and their descendants that the atomist theory was stored away in a musty intellectual drawer. The idea of a plurality of worlds went with it, where it remained more or less dormant in European thought for more than a thousand years.

What revived the idea of multiple worlds was not, as you might expect, the emergence of modern scientific thinking. It didn't come from Galileo or Newton or Copernicus. Instead, a burst of renewed interest came from the Catholic Church. The trigger was the rediscovery of Aristotle's great work *De caelo* (On the heavens). He'd written it before the dawn of Christianity, but the book wasn't translated into Latin until the late 1100s, long after the Church had become the dominant intellectual and social force in Europe. When *De caelo* reappeared, Catholic scholars—or simply, scholars, since there was virtually no other kind—scrutinized it to figure out whether the legendary philosopher's ideas were compatible with Church doctrine. Many of those ideas passed the test. The Earth, for example, was at the center of biblical creation, so Aristotle's putting it at the center of the world was perfectly appropriate.

But when Aristotle declared that other worlds couldn't exist, many believers thought he was going too far. Who was Aristotle, they asked, to say God couldn't make other worlds if he wanted? Multiple worlds might not arise naturally, acknowledged John Buridan, the rector of the University of Paris, in the early 1300s, but God could do things that wouldn't happen naturally. A half century later, another French scholar and cleric, Nicole Oresme, went a step further, arguing that other worlds *could* arise naturally. Aristotle's assertion that "down" must mean "toward the Earth" wasn't the only way to look at it. If you allowed "down" to mean "toward a heavy body," the problem went away. Leonardo da Vinci suggested that this might be the case for the Moon. The fact that the Moon doesn't fall to Earth suggests, he wrote in his notebooks, that it was its own center of attraction, with its own complement of water, air, and fire. The German theologian and philosopher Nicholas of Cusa, who influenced Leonardo's thinking, wrote:

> Rather than think that so many stars and parts of the heavens are uninhabited and that this earth of ours alone is peopled . . . we will suppose that in every region there are inhabitants, differing in nature by rank and all owing their origin to God, who is the centre and circumference of all stellar regions.

Even as theologians and scholars and natural philosophers were wrestling with Aristotle's ideas, his competition, the atomists, showed up as well, after a millennium in obscurity. The writings of the original atomists had disappeared with the rise of Aristotle, but a Roman philosopher named Lucretius, writing in the first century B.C., had preserved their ideas.

In 1417, Lucretius's atomist manuscript, titled *De rerum natura* (On the nature of things) was translated into Latin as well.

This was something of a problem. Having worked themselves into an outrage that some dead Greek had dared put a limit on how many worlds God could create, here came another dead Greek who rejected such a limit, but who was also pretty explicitly an atheist. Even worse, Lucretius's lyrical writing meant that his work was thought of primarily as a work of poetic literature at first. He sneaked onto everyone's must-read list before anyone could focus on his heresy.

Still, by the late 1500s, the notion of a plurality of worlds had become respectable, as long as it was expressed properly. God could make as many worlds as he wanted, in principle. But in practice, he'd just made ours (this was the line a young Galileo took, decades before he turned his telescope on the heavens). The arguments, as always, were purely theological, with no practical implications at all.

This wouldn't last long, however. In 1543, a Polish cleric and astronomer named Mikołaj Kopernik or Niklas Koppernigk, later Latinized to Nicolaus Copernicus, died and posthumously published a manuscript. Titled *De revolutionibus orbium caelestium* (On the revolutions of the heavenly spheres), it argued that the motions of the planets through the night sky could be best explained if the Earth and the other planets orbited the Sun, rather than everything orbiting the Earth. A preface to the manuscript suggested that this was a purely mathematical exercise—it didn't mean the Earth *actually* orbited the Sun. But a man named Andreas Osiander, who was overseeing the printing of the book, may well have added the preface without Copernicus's permission. Osiander was evidently appalled at the challenge to Church doctrine this new model of the solar system implied. The Earth must be central because it was in the very first sentence of the Bible, and because humans were the focus of God's closest and most loving attention.

As a result of both the preface and his death, Copernicus never got in trouble with Church authorities. The same can't be said for those who followed him—especially an ex-monk named Giordano Bruno, who was burned at the stake in 1600 by the Inquisition for a long list of heresies. Among them were his loudly proclaimed belief in magic, his rejection of orthodox Christianity in favor of a version incorporating Egyptian mysticism, his insistence that Copernicus's Sun-centered cosmology was *not* just a mathematical exercise, and his further insistence that other worlds weren't merely possible but they existed. "There are," he wrote in 1584 in the treatise *De l'infinito universo e mondi* (On the infinite universe and worlds),

> countless suns and countless earths all rotating round their suns in exactly the same way as the seven planets of our system [the Moon was considered a planet at the time]. We see only the suns because they are the largest bodies and are luminous, but their planets remain invisible to us because they are smaller and non-luminous. The countless worlds in the universe are no worse and no less inhabited than our Earth . . . Destroy the theories that the Earth is the center of the Universe!

In 1608, less than a decade after Bruno's execution, Galileo got wind of a Dutch invention called the perspective glass, which used lenses to make faraway objects look nearer. He built his own, improved version, which he called the perspicillum. The Dutch used their instrument to look at distant objects on the ground. Galileo, whose intellectual life was caught up in the heavens, turned his perspicillum in that direction as well. (The word *telescope* was coined in 1611 by the astronomer Giovanni Demisiani.)

Looking upward, Galileo immediately made a series of discoveries that made it impossible to think of the Copernican, Sun-centered solar system as just a mathematical fiction. He saw four unsuspected moons orbiting Jupiter, which meant they weren't orbiting the Earth. He saw moons around Saturn—or thought he did. The blurry bulges on either side of the planet that he believed were moons were actually Saturn's magnificent rings. He saw Venus go through phases, just like the Moon, proving that it, too, went around the Sun. He saw that our own Moon wasn't a smooth, perfect sphere, which was another of Aristotle's ideas the Church liked (things in the heavens must be perfect, they reasoned, because God is perfect). Giordano Bruno could reasonably be seen as something of a lunatic; all he had to offer were his beliefs. Galileo had evidence.

The Church resisted that evidence however. Finally, in 1633, the Inquisition forced Galileo to recant his belief in a Sun-centered universe and put him under house arrest for the rest of his life. But anyone who had access to a telescope could look through it and see everything Galileo had seen. Evidently, not everyone was willing to try. "My Dear Kepler," Galileo wrote to the German astronomer Johannes Kepler in 1610, "what do you have to say about the principal philosophers of this academy who are filled with the stubbornness of an asp and do not want to look at either the planets, the moon or the telescope, even though I have freely and deliberately offered them the opportunity a thousand times?"

Many did try, however, and for those in non-Catholic countries such as Germany and England, the Inquisition wasn't a threat. Within a century, thanks to Galileo, Kepler, Isaac Newton, John Flamsteed, Edmond Halley, and many others, the Earth had been all but completely displaced from the center of the universe in Western thinking. Thanks to Newton in particular, people who studied the skies now understood that the same physical laws applied in the heavens and on the Earth.

But Kepler, especially, had taken things much further. The Moon, he argued, wasn't just superficially like the Earth in being spherical and mountainous. Even before the telescope had been invented, he was arguing that the Moon also had an atmosphere and weather. He also suspected—surprisingly, given that the idea of extraterrestrial life feels relatively modern—that it was inhabited. Even more surprisingly, he wasn't the first. As far back as the sixth century B.C., the Pythagorean school of philosophy held (according, at least, to secondhand writings) that "the moon is . . . inhabited as our earth is, and contains animals of a larger size and plants of a rarer beauty than our globe affords. The animals in their virtue and energy are fifteen degrees superior to ours, emit nothing excrementitious, and the days are fifteen times longer."

Kepler himself imagined two races of Moon people in his 1634 work *Somnium* (The dream), a fictional account of a trip to the Moon. The Subvolvans lived on the familiar side of the Moon that faces the Earth; the Privolvans lived on the far side. But it wasn't just the Moon that was inhabited: Kepler also concluded there must be intelligent beings on Jupiter. The moons Galileo had found proved it: Since they weren't visible to the naked eye from Earth, God must have created them for the Jovians to look at.

A century and a half later, William Herschel—the German musician who taught himself astronomy in middle age and went on to build the finest telescopes in existence, find the planet Uranus, discover infrared radiation, and much more— was a firm believer in extraterrestrials. The Moon was certainly populated, he was sure. In a letter to a friend in the Royal Society he wrote, "If you promise not to call me a Lunatic, I will . . . shew my real sentiments on the subject." So was Mars, which he wrote, has an atmosphere, "so that its inhabitants enjoy a situation in many respects similar to ours." So was the Sun. The people who live there don't burn up, he believed, because the fiery surface we see is actually the blazing top of a permanent cloud cover. The undersides of these burning clouds are cool, and the dark sunspots we see are breaks in the overcast, giving us a glimpse of the cool and comfortable surface that lies below.

While Herschel and many others populated just the solar system with their speculations, followers of the French philosopher

René Descartes, who did most of his work in the mid-1600s, theorized on a grander scale. Descartes rightly guessed that the stars were faraway suns in their own right. He stopped there, but one of his students, Bernard Le Bouvier de Fontenelle, took the next logical step. "Our Sun enlightens the Planets; why may not every fix'd Star have Planets to which they give light?" he asked the French in his 1686 book *Entretiens sur la pluralité des mondes* (Conversations on the plurality of worlds).

And so, from the early 1600s on, philosophers and scientists would debate the questions that would ultimately lead Frank Drake, the founder of SETI, to begin listening for alien signals in 1961 and would lead Geoff Marcy and Bill Borucki to begin searching for extrasolar planets in the mid-1980s. They couldn't test their theories, though. Despite the construction of more and more powerful telescopes through the late 1800s and early 1900s, there was no way to back up their ideas with observations. Planetary formation could, depending on whose theory you liked, be inevitable or improbable. Our solar system might be typical, or very rare.

With those same telescopes, however, astronomers kept finding new objects in the solar system—scores of asteroids starting in 1800; the planet Neptune in 1846; moons circling Uranus, Mars, Jupiter, Saturn, and Neptune throughout the 1800s; Pluto in 1930. Throughout most of that period, scientists became more and more convinced that at least some planets were populated—especially Mars and Venus—although Herschel's idea of creatures that walked the Sun never caught on.

By the late 1800s, almost all of that conviction was focused on Mars. Venus was perpetually covered with a thick overcast that surrounded the entire planet. You could speculate about a dank, swampy surface, and many did, but speculation was the end of it. With Mars, by contrast, the surface was visible, in a blurry sort of way. Swimming in and out of view in that blur were two very notable features. First, while most of the surface was a reddish orange, the Martian poles were white, suggesting that Mars had icecaps, like those on Earth. The observers could even see the white spots wax and wane with the Martian seasons. That proved Mars had an atmosphere that could move melted, evaporated ice from one hemisphere to the other to fall as snow each winter.

But that was nothing compared with the evidence first reported by Italian astronomer Giovanni Schiaparelli in 1877. At the very edge of visibility, Schiaparelli said, he saw markings on the Martian surface. In an article published several years later, he wrote,

> All the vast extent of the continents is furrowed upon every side by a network of numerous lines or fine stripes of a more or less pronounced dark color, whose aspect is very variable. These traverse the planet for long distances in regular lines, that do not at all resemble the winding courses of our streams.

Not only that: These *canali*, or channels, seemed to widen and narrow with the change of seasons, and some of them appeared to double into pairs of parallel channels. He continued,

Their singular aspect, and their being drawn with absolute geometrical precision, as if they were the work of rule or compass, has led some to see in them the work of intelligent beings, inhabitants of the planet. I am very careful not to combat this supposition, which includes nothing impossible.

In many of his scientific papers, Schiaparelli was careful to acknowledge that much of what he saw could also be completely natural, and that his eyes might even be playing tricks on him. Some of the other astronomers who also saw the *canali* were convinced that they were, in fact, optical illusions, but others were sure they were real. In other writings, Schiaparelli made it clear that he strongly favored the argument that many of the canals were engineered by Martians. The thickening, he believed, was a seasonal expansion of vegetation as the *canali* flushed with water and nourished crops on either side.

Schiaparelli was a model of caution and restraint, however, compared with Percival Lowell. Lowell, a rich Bostonian and the brother of future Harvard president A. Lawrence Lowell, became intrigued with Mars in the early 1890s. He built his own observatory under the clear skies of Flagstaff, Arizona, equipped it with a much more powerful telescope than Schiaparelli used, and began scanning Mars for evidence of the life he was already convinced was there. With that attitude, it's no surprise that he found it. "Mr. Lowell went direct from the lecture hall to his observatory in Arizona," wrote astronomer W. W. Campbell disparagingly in a review of Lowell's popular

1895 bestseller *Mars*, "and how well his observations established his pre-observational views is told in this book."

This end run around the scientific method didn't bother the press, however. MARS INHABITED, SAYS PROF. LOWELL, ran a headline in the *New York Times* on August 30, 1907. "Declares the Planet to Be the Abode of Intelligent, Constructive Life . . . Changes in Canals Confirm Former Theory—Splendid Photographs Were Obtained." An editorial in the *Times* from around the same time says in part, "Harvard has ignored Prof. Lowell's discoveries of water vapor, vegetation, snow caps, and canals, always, more canals, on Mars. The people of this country support Prof. Lowell in his Martian campaign."

Lowell died in 1916, still convinced he had discovered intelligent life on Mars. But within a few years after his death, the canals had vanished, as bigger and more powerful telescopes gave astronomers a sharper view of the planet. They were, as Schiaparelli had admitted might be the case (but Lowell hadn't), a trick of light, an optical illusion. For the general public, however, which had been reading about scientific claims of Martian life for decades, the idea of living Martians had become part of the popular culture.

It wasn't just public lectures and glowing newspaper stories. In 1898, H. G. Wells published the bestselling *War of the Worlds*. In 1912, Edgar Rice Burroughs, who a year earlier had been working as a wholesale distributor of pencil sharpeners, published *A Princess of Mars*, followed by nearly a dozen sequels. He went on to write a series about life on Venus as well, and one novel titled *Skeleton Men of Jupiter*. (Burroughs also

invented Tarzan of the Apes, by far his most famous charac-
ter.) By the time Orson Welles turned *War of the Worlds* into a
radio play in 1938, the idea of intelligent Martians was so
plausible that some listeners thought an invasion was really
going on—although the idea that hundreds of thousands of
Americans fled their homes in panic is evidently an urban leg-
end. Most of those who had listened to Welles's program the
week before didn't panic, since he told them to stay tuned next
time for the Martian invasion.

Even as aliens became a staple of B movies and TV shows
such as *The Twilight Zone* and of breathless reports about UFOs
in the 1950s—and maybe partly as a result—the topic moved
out to the scientific fringe. The physicists Philip Morrison and
Giuseppe Cocconi went out on a limb when they published a
paper in 1959 titled "Searching for Interstellar Communications," which suggested listening for alien radio signals. Morrison and Cocconi were no Percival Lowells; they didn't claim
that extraterrestrials existed. "The probability of success is
difficult to estimate," they wrote at the conclusion of the
paper. "But if we never search, the chance of success is zero."
Frank Drake independently began his search at the National
Radio Astronomy Observatory in Green Bank, West Virginia—
but he kept it a secret. The search for alien signals is still
going on today, but without a lot of funding, and with the
understanding that the search still has to overcome what astronomers call the "giggle factor."

By the early 1960s, the Soviet Union had beaten America
into space with the unmanned *Sputnik 1* satellite, and again
with the first orbital flight by cosmonaut Yuri Gagarin. In a

Cold War panic, John F. Kennedy declared that America would demonstrate its superiority by beating the Soviets to the Moon. Money poured into the U.S. space program, which, despite waving flags and inspiring words about exploration and frontiers and understanding the universe, was mostly about showing the United States was better than the Communist enemy. Infused with money, a mandate to reclaim national pride, and a swelling staff of young scientists and engineers, NASA began shooting American astronauts into space and sending probes to the Moon, scouting for landing spots. The agency began sending robotic spacecraft out to explore the solar system as well—inward to Mars and Venus and out toward Jupiter and Saturn.

Whatever the original motivation, the *Mariners* and *Pioneers* and *Voyagers* brought back spectacular images and all sorts of data about the planets and their moons. The information, along with other observations, pretty much destroyed any remaining speculation about life on most of them. Venus turned out to be so hot that lead would melt on its surface. Mercury was covered with craters and relentlessly scorched by the nearby Sun. Jupiter, Saturn, Uranus, and Neptune were already known to have no surface at all until you got thousands of miles down into their gaseous atmospheres. Pluto, still a planet back then, was too frigid. But Mars proved to have ancient riverbeds; it clearly had no farmers or engineers, but it might once have had some primitive form of life, billions of years ago. The scientific search for life on other planets in our own solar system was revived, refocused on microbes rather than Martian princesses or skeleton men from Jupiter.

Thanks to more powerful telescopes than Galileo or Kepler or Herschel could have imagined, meanwhile, the active search for planets orbiting other stars, and not just theoretical speculations about them, was revived. NASA would try to play a major role here as well. But much of the agency's effort toward planet-searching would end up being wasted.

Chapter 5

THE DWARF-STAR STRATEGY

D AVID CHARBONNEAU ARRIVED at Harvard in the
fall of 1996, ready to literally take on the universe. He'd
majored in physics as an undergraduate at the University of
Toronto, concentrating on the physics of the cosmos just after
the Big Bang. Thirty years earlier, a couple of radio astronomers at Bell Laboratories in New Jersey had stumbled on a
mysterious whisper of static coming from all directions in the
sky at once. The static turned out to be the feeble remains of
an ancient flash of light from a time, more than thirteen billion
years ago, when the universe emerged from a state of unimaginable density and heat and began to condense into the galaxies and stars we see today. The light was originally white-hot,
but as the universe expanded, it cooled to red, and then moved
out of the visible range entirely and into the microwave part
of the electromagnetic spectrum, more of a faint glow by now
than a flash.

Physicists realized that hidden within this cosmic glow—
they named it the cosmic microwave background, or CMB—
must be all sorts of information about the early universe. They

presumed, although they couldn't yet see them, that some spots should be a little warmer than average, representing spots of slightly higher than average density in the baby universe. These dense regions would have eventually formed into clusters of galaxies, presumably, while cool spots, with unusually low density, would have become the empty spaces in between. The warm and cool spots turned out to be so hard to detect that it took until 1992, and even then, the Cosmic Background Explorer (COBE) satellite that saw them could capture only the blurriest of images. If the spots hadn't shown up at all, it would have been a serious problem for the Big Bang theory. But while it confirmed the Big Bang, the data COBE sent back to Earth wasn't enough.

By the time Charbonneau was a senior at Toronto, physicists and astrophysicists had started to do follow-up studies from the ground to refine their understanding of the CMB, and they were starting to design new satellites that would improve on COBE. Toronto's Canadian Institute for Theoretical Astrophysics was in the thick of this, led by a prominent theorist named Dick Bond who was trying to figure out what the emerging patterns of hot and cold spots meant. "There were all of these wild data sets coming out of Antarctica," Charbonneau told me one afternoon in 2010, "and from Saskatoon, in Saskatchewan. I was feeding off the excitement of the cosmologists."

Charbonneau is now a full professor at Harvard, and our conversation took place as we walked from the Harvard Science Center, where he'd just finished teaching a class, up to the Center for Astrophysics a half mile or so away. It wasn't a

Dave Charbonneau (Stephanie Mitchell/Harvard staff photographer)

leisurely walk. Charbonneau is extremely tall, so his stride covers a lot of ground. He's also extremely busy. He's got his own very time-consuming research project, he's a member of the Kepler science team, and he's also a very popular teacher. "Normally," he told me, "I only teach one class, but this semester I've elected to teach three, because I really care about our undergraduate program."

The world of astrophysics was so consumed with cosmology in the mid-1990s that even though the first planets found by Mayor and Marcy and the rest were announced during his senior year in college, Charbonneau doesn't remember hearing a word about it. "It was just so new," he said, "and it wasn't percolating down to the undergraduates. I'm sure the graduate students knew about it, but what we teach undergrads is at least five years out of date in terms of excitement." In fact, he recalled, "I was doing my fourth-year course in mathematical physics at the time—beautiful stuff, but the content hasn't changed in a hundred years."

When he arrived in Cambridge as a graduate student, Charbonneau fully intended to stay with cosmology. The faculty had put together a series of afternoon seminars, though, in which professors were talking about new exciting things that they thought the grad students might want to get involved in. One of those new exciting things was exoplanets. The Mayor and Marcy discoveries had quickly caught the attention of astronomers around the world. Now that everyone knew you could actually find planets, it seemed that everyone wanted to get in on the action. "There was a guy named Bob Noyes," said Charbonneau, "and he got up at one of these afternoon seminars and gave this talk about how there was this big debate as to whether these really were planets."

In part, the doubt centered on the same problem that had faced Dave Latham and his collaborators back in 1989. If an object's orbit around a distant star lies edge-on from the perspective of Earth, then you're seeing 100 percent of the wobble it causes in the star. If it's not edge-on, then some of that wobble is going in a direction you can't detect. You'll see only 90 percent, or 80 percent, or even less, which means you'll underestimate the mass of the orbiting object. Marcy and Mayor were always careful to talk about their exoplanets' *minimum* masses—the mass the planets would have if they were truly orbiting edge-on. If the handful of exoplanets known at the time had a lot more than the minimum mass, which was certainly possible, they were too massive to be true planets. Another possibility, as Marcy had anticipated, was that the stars themselves were pulsing; if so, maybe it was just the surface, not the entire star, that was moving toward and away from the telescope.

Marcy had good arguments for why both of those scenarios were unlikely, but scientists can be awfully clever at coming up with good arguments. If someone could confirm the existence of his planets and Mayor's in an independent way, it would all be a lot more convincing. Noyes, a senior faculty member whose office was right across the hall from Charbonneau's, thought he had a way to do it. If the planets really were orbiting edge-on, they should show phases, like the Moon—or, more aptly, like the phases of Venus, which Galileo discovered. When Venus is between the Sun and the Earth, we see its unlit side. When it's on the other side of the Sun, we see its fully illuminated side (every so often, Venus is *exactly* on the other side of the Sun, so we don't see it at all, but this happens rarely). The same should be true for an exoplanet: It should be dark when it's on the near side of its orbit and bright when it's on the far side. You can't actually see the planet; it's much too small, and too close to the glare of the star. But you can see the total amount of light coming from the star and planet together. So when the planet is bright, the sum of star plus planet should be a little bit brighter; when the planet is dark, the sum should be a little dimmer.

This is what Noyes proposed to look for, and when he

explained it, Charbonneau was hooked. "One thing that appealed to me," he recalled, "is that the cosmology stuff is so amorphous. I mean, if you have to give a public lecture about why people should care about the CMB, you're crippled by the fact that the main thing you want to talk about is just not accessible to someone who doesn't understand spherical harmonics." Exoplanets, he realized, were completely different. "The basic idea is something that I can explain to anyone in just a few sentences," he said, "and it doesn't require any mathematics. People can visualize what you're talking about."

Beyond that, as a grad student in cosmology he would be working on problems that some of the most accomplished scientists in the world had already been wrestling with for decades. "I had this feeling that to get going in cosmology I had to read ten or twenty years' worth of papers before I could begin work," he said. That wasn't true for exoplanets. "Literally, I had one folder and as each paper came out on exoplanets, it would go in that folder. There was a period of time when there were basically five or ten papers that you had to have read."

In short, he realized, it was one of those rare moments in scientific history when a brand-new field opens up and graduate students can identify and answer very simple questions and make a real contribution. You could also get research grants relatively easily for exoplanet research. "Normally, when you write a proposal for money or for telescope time, you spend most of the time explaining why what you're doing is interesting. But we never had to explain why it was interesting to go and look for new planets. It was obvious."

Charbonneau was excited by the idea of working on exo-planets for a more fundamental reason as well. "It's life—the idea of searching for life in the universe. It used to be the case that professors were very hesitant to say that, so they would always say, 'It's the physics of planet formation,' and that's as far as they would go. But I'm not afraid to say that it is absolutely this question of 'Are there examples of life that arose indepen-dently from the life on the Earth?'" He was even willing to say it back in the mid-1990s, when all the known exoplanets were much too large to support life, or much too hot, or both. By the time we spoke in 2010, it was a more realistic question to ask.

"So," he explained, "you basically have a mix, you have a ball of silicate rock, you put an ocean down, not too thick, and you have roughly the right temperature and atmosphere, then is it just a matter of time when life will arise?" Or would it come out differently? "Are we going to find, say ten or twenty examples of those"—a proposition that was looking better all the time, with Kepler already in orbit for more than a year and with at least one planet plausibly made of rock, not gas, dis-covered by Michel Mayor's group a year or two earlier— "study them carefully, find that, although they look just like the Earth in terms of their properties, they just don't have life for whatever reason?"

For all his excitement at the prospect of life, Charbonneau thinks either outcome is equally possible. "People push me on this," he said, "and I really, honest to God, think it could go either way. When I teach, most of my students can't imagine the latter case. They think that of course there will be life."

That doesn't mean for a moment that he's indifferent to the answer. He hopes life exists beyond Earth. "Even if they were completely foreign, I think we would feel less lonely, I think there would be this true loneliness if we found out that this was really it. And I think that does affect how people view how precious our planet is. Even though we could never go to those other places, I think that we would still view what we have differently if we knew that it was truly unique."

These insights weren't quite so fully formed, of course, when Charbonneau initially switched from cosmology to planet-hunting. But Bob Noyes's scheme of looking for reflected light was instantly appealing. "I worked very hard on that," said Charbonneau, "but we never did make a detection. It turns out that reflected light is a really hard problem. I always felt we were almost there, but it took Kepler to pull it off, six hundred million dollars and fifteen years later." After this disappointing finish to the reflected-light project he went back to Noyes, who had become his adviser, to get advice on a research topic for his thesis, the grand finale to a graduate student's career. "Bob told me it probably wasn't the best bet to go into radial-velocity searches [like Marcy and Mayor were doing] because so many teams were so far ahead," recalled Charbonneau. "Maybe I could try to confirm planets by looking for transits." This made a lot of sense: It's hard to imagine something other than a planet that could make a star wobble *and* make it dim, with exactly the same timing. Even so, while it was now 1999, at least a decade and a half after Bill Borucki had begun working seriously on what would become the Kepler mission, planetary transits were still something very few other astronomers were thinking about.

Confirming that a planet really existed was one reason to look for transits, but there was an even better one, which wasn't fully appreciated at the time. When Marcy and Mayor found the radial-velocity signature of an orbiting planet, they could tell you how much time the planet took to complete one orbit, and they could tell you its mass—or its minimum mass, anyway. If you could see that a planet made a transit as well, that would confirm you were seeing the orbit precisely edge-on, so you would know that the minimum mass was also the actual mass. But you would also know the planet's physical size. If it blocked, say, 2 percent of the star's light, that meant the planet's disk was 2 percent of the size of the star's disk. With a little bit of high school geometry, you could calculate the planet's volume. And once you had the mass and the volume, you knew the planet's density.

Planetary scientists already know from our own solar system that planets come in different densities. Mercury, for example, is denser than Earth because it has a higher proportion of iron to rock than Earth does. Pluto, whether you call it a planet or a dwarf planet, is less dense than Earth because it's made largely of ice. Saturn is even less dense, because most of its mass comes in the form of hydrogen and other gases, with only a little bit of rock down in its core. One of astronomy's longest-standing fun facts is that if you could find a big enough bucket of water to put it in, Saturn would float.

Whether you're interested in finding a place where life

might exist, or simply in understanding how an alien planet or solar system formed, it's crucial to know what the planet is made of. A planet with the mass of Earth bloated out to the size of Jupiter, to take an extreme and, in fact, physically impossible example, would be a wispy ball of dilute gas—not a good place to live. A planet the size of Earth but with a mass fifteen times as high would be so dense, and have such a crushing surface gravity, that probably nothing could live there either. For all of these reasons, the detection of planets where astronomers already had radial velocities would be incredibly valuable.

In 1999, when Dave Charbonneau was casting about for a thesis topic, Bill Borucki hadn't yet gotten approval for Kepler. But a handful of other astronomers had begun thinking about transits. Tim Brown was not only thinking about them; he was actively searching. Brown was originally interested in studying the Sun, and by the 1980s was working at the National Center for Atmospheric Research (NCAR), in Boulder, Colorado. NCAR, which mostly concerns itself with climate, isn't exactly a hotbed of astronomy, but since the Sun has a big effect on Earth's climate, understanding how it works is perfectly in keeping with the lab's mission. To give just one example, the Sun gets a little brighter overall when it has lots of sunspots, and dimmer when they almost disappear, in a regular boom-and-bust eleven-year cycle. During the 1600s and 1700s, though, sunspots pretty much disappeared entirely for an eighty-year stretch. At the same time, Europe experienced a period of unusual cold, known as the Little Ice

Age. There's good reason to think the cold spell was largely due to other factors, but even so, it's useful to try to figure out how this and other changes in the Sun might affect the Earth— especially since some solar physicists think the Sun may now be entering another prolonged sunspot drought.

While he was at NCAR, Brown built a spectrograph to measure subtle pulsations in the solar surface so he could figure out what was going on inside. It could measure pulsations in other stars as well, and Brown realized that he might be able to use spectrography to look for the wobbles caused by orbiting planets. "Planets were always sort of an afterthought," he told me during a telephone conversation during the summer of 2011. But he was aware that Geoff Marcy was looking for planets, and he knew about Michel Mayor and the Swiss group, and about the group in Canada. By the mid-1990s Brown had teamed up with a few other astronomers including Bob Noyes at the Harvard-Smithsonian Center for Astrophysics, or CfA—the umbrella organization, located within a single building complex in Cambridge, that includes the Harvard University astronomy department and the Smithsonian Astrophysical Observatory.

The Brown-Noyes group set up shop at the Smithsonian's sixty-inch telescope on Mt. Hopkins, near Tucson, Arizona. They were mostly looking for pulsing stars, but they kept their eyes out for planets as well. "It's fair to say that we didn't get anywhere," Brown told me in a phone conversation, "but it's a curiosity that I do have two observations of 51 Pegasi that predated Michel Mayor's discovery." He made them in 1995, about six months before Mayor announced he'd found a planet.

Brown looked at the data, saw what turned out to be the signature of 51 Peg b, and thought, "Aha, the spectrograph is probably misbehaving again." After that, the team dabbled a bit more in looking for planet-induced wobbles, but, said Brown, "we were never seriously in that game, although we tried to be."

But as early as 1992, he and his frequent collaborator Ron Gilliland, of the Space Telescope Science Institute, had already begun thinking about looking for planets by means of transits. It was mostly theoretical at first, since everyone still thought that other solar systems would resemble ours. If that was true, then the biggest, most easily detectable planets (like Jupiter) circle their stars in long, loping orbits. If they did transit, it would be only once every decade or more. Even then, the odds were long that they could spot such a transit: At a distance of hundreds of millions of miles from the star, the plane of a planet's orbit would have to be so precisely aligned with our line of sight that such a thing would happen only rarely. A huge planet orbiting right up against its star wouldn't have to be aligned with nearly so much precision—but such a planet was beyond the imagination of most astronomers at the time. "But then in '95," said Brown, "when Mayor found 51 Peg in such a tight orbit, it became obvious that it was sensible to go looking. It was obvious to me, anyway. I had trouble convincing others."

He and Gilliland had already convinced themselves, however, that they could achieve the necessary precision to detect transits, even without a huge telescope. Brown and Gilliland got together with Ted Dunham at Lowell Observatory—the same place where Percival Lowell had found "proof" that Mars was inhabited—to cobble together their first transit-search telescope. (Dunham would ultimately become the Science Team director for the Kepler Mission before leaving to work on a high-altitude infrared observatory.) By 1999, they had it built, and Brown set it up in a parking lot at NCAR to run it through some tests.

It was just at this moment when Bob Noyes, Brown's old collaborator, was trying to help Dave Charbonneau find a thesis topic. Noyes knew about the transit-search project (Brown called it STARE, for STellar Astrophysics and Research on Exoplanets), and he suggested that Charbonneau head west to help out. "I bought a car," recalled Charbonneau, "and I drove out to Boulder, and started working with Tim." Before he went, he stopped to see Dave Latham. As Latham remembers it, "Dave ambushed me outside the door to my office as I was leaving one night and said, 'I'm leaving for Boulder, and I need your advice on some good objects to look at.'"

Although Latham had let Michel Mayor and Tsevi Mazeh pick up the planet-hunting project and run with it a full decade earlier, he was still in touch with them. Latham knew, although it hadn't been announced publicly, that Mayor and Mazeh had found a radial-velocity signal in a star called HD 209458. The size of the planet, and the nature of its orbit—assuming it was a planet, naturally—made it a prime candidate to transit across the face of the star. "Look at HD 209458," Latham advised. "I'll tell you exactly when to look." This was in late August. Charbonneau forwarded Latham's heads-up to Colorado, where, sure enough, Brown found the first transit. "I think it was 9/9/99, September 9, 1999," said Charbonneau, as we walked from his class to his office. "That was the first recorded transit of an exoplanet in front of its star." Charbonneau showed up in Boulder a couple of days later, in time for the second transit, on September 16.

But Charbonneau and Brown figured this out only in retrospect, a couple of months later. That's because Brown had another search for transits going on at the same time, with his old partner Ron Gilliland. This one used the Hubble Space Telescope, in a sort of preview of the Kepler Mission. As Bill Borucki had realized ten years earlier, the chance of seeing a transit on a single random star (that is, one where Dave Latham hasn't tipped you off first) is vanishingly small. The best way to search is to look at a field of stars all at once. With its small field of view, Hubble isn't the ideal instrument to use, so Gilliland and Brown cheated a little: They focused on one of the hundreds of globular clusters that dot the central regions of the Milky Way.

Globular clusters are knots of up to a million stars—almost like miniature, spherical galaxies. The astronomers pointed the Hubble at a cluster called 47 Tucanae, and took data on about thirty-five thousand stars. 47 Tuc was much too far away for radial-velocity measurements, so there was no hope of getting densities for any planets they might find—but getting a sense of how many planets there might be in this big sample, and what sizes and orbits they came in, would still be useful. They predicted they might find seventeen transits. In the end, they found zero, perhaps because globular clusters contain mostly very old stars that have relatively little of the elements planets are made of.

Brown and Gilliland had gotten what Brown calls "a big pile of Space Telescope data" just as he and Charbonneau began doing their ground-based observations from the back parking lot. So they put aside the STARE data and began working their way through the Hubble data. Before they could get very far, Charbonneau got a call from back home. It was John Huchra, the director of graduate studies and former head of the Harvard astronomy department. Huchra was an unrepentant cosmologist—none of this planet stuff for him!—who had helped create the first large-scale maps of the universe in the eighties.

He was calling because Charbonneau had never gotten official permission to go west. Bob Noyes knew about the trip, but that wasn't good enough. "Huchra said, 'Dave, you've got to come back and you have to make a case for the science you're doing. We're really nervous about students going away and working with scientists outside of the CfA because we can't supervise you.' I think maybe he thought I'd gone out to Colorado to go skiing, I don't know," Charbonneau told me.

In any case, he rushed back to Cambridge. "I remember I went through a long defense—it was a couple of hours, at least." About ten minutes of that was talking about why transits would be interesting; that part was pretty obvious. "The rest," he said, "was what my thesis would say when I didn't find a transiting planet, which was what they all assumed would happen." So he spent the rest of the time talking about the things they believed he *would* find, like binary stars and pulsating stars, and all the good science he could wring out of those. "Of course," Charbonneau continued, "the irony is that I already had the data that would show the transits of 209458, even though I didn't know it yet."

The thesis committee was satisfied and Charbonneau headed back to Boulder. In the end, it took two months of working on the Hubble data before they could get back to their parking lot project. At almost exactly the same time, though, Brown got wind that Geoff Marcy and a collaborator named Greg Henry, at Tennessee State, had detected a transit as well, although he didn't know what star they were observing. "Geoff and I had an interesting phone conversation," he recalled, "which amounted basically to 'You show me yours and I'll show you mine.'" It turned out to be HD 209458. Brown and Charbonneau didn't get extra credit for the fact that their observations beat Marcy and Henry's by two months, since they hadn't analyzed them. The two teams published their discoveries in the same issue of the *Astrophysical Journal* the following year, and they're generally given equal credit for the discovery (although some websites mention just one or the other; a news release published on the UC Berkeley website at the time, for example, mentions only Marcy's group).

Credit aside, though, the observations marked another crucial step forward in the search for Earth-like planets. Four years earlier, Michel Mayor and Geoff Marcy and their teams had made a strong case that planets, or something that looked a lot like planets, orbited around Sun-like stars. With the discovery that at least one set of radial-velocity wobbles was matched by a series of transits, with precisely the same timing, it became impossible to doubt that many of them, if not most, had to be planets.

Moreover, the fact that the planet now known as HD 209458 b already had a known mass, thanks to Mayor and Mazeh, and that its size was now known from the amount of light it blocked during its transits, allowed the astronomers to calculate its density. It was surprisingly low: HD 209458 b turned out to be about 70 percent as massive as Jupiter, but about 35 percent bigger. With such a low density, it was clearly made mostly of gases, but what kinds, and in what percentages, and with what implications for the makeup of smaller, life-friendly worlds, were still unknown.

Before we finished our conversation, Charbonneau wanted to tell me one more thing. "When you defend your thesis," he said, "you have to have an outside reader, in addition to members of your department. I picked Bill Borucki." At the time, very few of the astrophysicists at Harvard knew much about Borucki, a government scientist who didn't even have a Ph.D. But Charbonneau did. "I already knew that he was going to be PI of the Kepler Mission and I knew Kepler would find all of these planets, these Earth-like planets. Hopefully, anyway."

Chapter 6

IMAGINING ALIEN ATMOSPHERES

W HEN DAVE CHARBONNEAU arrived at Harvard, it turned out that he wasn't the first Canadian from the University of Toronto to show up there intending to be a cosmologist. Two years earlier, a woman named Sara Seager had come to Harvard with exactly the same intention. They'd known each other back in Toronto, but she had been two years ahead. "That doesn't seem like a lot now, but as you know, when you're younger it's a much bigger deal," she told me one morning at her office at MIT. No one would ever describe the MIT campus as charming; it's stark and soulless compared with leafy, ivy-covered, Georgian-brick Harvard a couple miles to the northwest. But Seager's office is on an upper floor in the Green Building—a high-rise that is typically bleak on the outside, but with a view from the higher floors that plenty of Harvard professors would undoubtedly kill for. Her window looks down on the Charles River Basin, filled with sailboats and rowing shells for much of the spring, summer, and fall. Across the basin, the towers of downtown Boston loom only a mile or so away. Seager and Charbonneau had first met, she told me, when she was a senior and Charbonneau was a sophomore. "He was thinking of dropping out of physics," she said, "and after I graduated he sent me this really nice letter thanking me for convincing him not to."

When she first got to Harvard, Seager worked on the recombination of the universe. This is the time, about four hundred thousand years after the Big Bang, when the hot, dense universe cooled off enough for atoms to form out of subatomic particles. The light that burst free as a result of that event is what astronomers stumbled on in 1965, proving the Big Bang had happened, and what cosmologists use to try to figure out how matter in the universe was arranged at the time. Like Charbonneau, Seager was working on the most popular topic in astrophysics in the mid-1990s—just as it was starting, like the early universe itself, to cool off. "My most highly cited papers are still the ones I wrote in cosmology," she said. "But that's because there are more cosmologists to cite papers, just so you know."

This is how a conversation with Sara Seager often goes. She's charming and gracious (which is a pretty generic quality, it turns out, among planet hunters), but she talks with a palpable intensity and focus. Her mind always seems to be in high gear, and she sometimes drops in an observation, like the one about cosmologists citing papers, that gives a satisfying flash of insight—but only for a millisecond, because by then she's already moved on to another point I don't want to miss. It happened again one time when she was talking about her father, who sparked her interest in astronomy when he took her to an amateur star party as a young girl.

"I remember looking at the Moon through a telescope," she said. "It was unbelievable. Do you remember the first time you ever saw the Moon through a telescope? Do you remember seeing it recently? Or ever? Isn't it so incredible? It is just unbelievable."

Anyway, she went on, her father was a doctor in a suburb of Toronto, and he got frustrated with the Canadian medical system, which kept exerting more and more control about what kinds of procedures you could do, and how much you could charge. So, she told me, he decided to leave medicine and go into the hair-transplant business. "He always wanted to have hair," she told me, "because he lost all his hair by the time he was nineteen, and people told him he couldn't have it. You know," she said, in one of her characteristic asides, "most men who have no hair don't mind. Only one percent of the men without hair wish they had it." I have no idea whether this is actually true, but knowing Sara Seager I'm fairly certain she looked it up once, and never forgot it.

Seager's biggest contribution to cosmology is a detailed calculation of how the physics of recombination must have unfolded—a theoretical analysis that would help observers make sense of what the satellites were seeing. "So I finished that; it was done. So what next?" Her thesis adviser, Dimitar Sasselov, had been at the conference in Florence where Michel Mayor announced the discovery of 51 Pegasi b, and he'd been sufficiently excited that he had started to move from cosmology into exoplanet theory himself. "Of course, we were talking about that," Sasselov told me during a visit to Cambridge many years later, "and Sara said, 'You know, I like this

Sara Seager (Justin Knight)

cosmology stuff, but I really want to work on planets.' So we came up with a project for her." At the time, Marcy and Mayor between them had found only a handful of planets, and three or four were "hot Jupiters" like 51 Peg b. The atmospheres of these planets were hot gases. The early universe was a hot gas. That meant Seager could apply some of her cosmology calculations to planetary atmospheres.

Some of the faculty on her thesis committee thought that the planets didn't even exist and that nobody would be able to observe them and test her theoretical predictions in any case. "But Bob Noyes was also on my thesis committee," Seager told me, "and he really believed in these planets. He was like, 'Don't listen to these people, this is going to be good.'" For the next two years, Seager and Sasselov worked together on the atmospheres of hot Jupiters and on the early universe, both at the same time. "I'm sure nobody does this anymore," Sasselov told me, "and from the perspective of ten years or more of time, it feels like we jumped from one field to another in a sort of instantaneous split. But if you look at it with the resolution of a month, you can see that it didn't really happen instantaneously. You're in one field, and then, at some point, you cross into the other field. I don't work on the early universe anymore."

As she went ahead and worked out her thesis, the same insight came to Seager that had come to Tim Brown. If these hot Jupiters really existed, they were hugging their stars so tightly that at least some of them should produce transits. "In retrospect," she said, "one of the things I was really good at was having hunches about things that were really going to happen.

So now I try to take action on those hunches because I am older and wiser." At the time, Dave Latham was working in an office right nearby, and, recalled Seager, "every time I saw him I was like, 'Dave, you've got to look for transits, transits are about to happen.' I knew he had a few radial-velocity candidates and I kept telling him over and over again, because it was fresh on my mind, 'You've got to do this. I'm sure you must have a transit in there.'"

Latham remembers things the same way, except he remembers Dave Charbonneau bugging him as well. And Charbonneau remembers getting the idea from Bob Noyes. Trial lawyers know that eyewitnesses to an event often remember vivid details—but that the details don't match up with the memories of other witnesses. Whatever the case, it's clear that the idea of looking for transits was in the air at Harvard in 1998 and 1999. Maybe Seager thought of it first, or maybe it was Bob Noyes, or maybe it just emerged out of multiple conversations between different pairs of astronomers, without anyone deserving full credit. What happened next is that Seager, her thesis finished and approved, headed off to the Institute for Advanced Study, in Princeton, New Jersey, to take a five-year postdoctoral fellowship. She was still a theorist, not an observer, but she couldn't stop thinking about transits.

"When I went to Princeton," she said, "I tried to figure out if I could make an observation myself. But do you know how many photometric nights they get a year in New Jersey?" she asked, referring to the kind of clear, dry, Moonless nights when you can make high-precision measurements of an object's brightness. "They get one per year. I mean, come on, it's hazy all the time there." Besides, Princeton's only good-size telescope didn't have the light detectors she needed to make that kind of observation. She checked back at the University of Toronto, but they didn't have the right equipment either. "So after that, I dropped it. It wasn't really my thing anyway."

This was in the fall of 1999. The following January, she was at a conference where she ran into Dave Latham. He gave her a huge hug. "I mean, people don't usually give hugs in astrophysics. It's not protocol. And I was like, 'What's going on?' And he told me he was just so happy I'd bugged him about transits, because Dave Charbonneau, who is one of the luckiest people on Earth, had just found the first transiting planet. So I was like, 'Oh, my God, I could have found the first transiting planet, but I just didn't pursue it enough.'"

Seager was hired at the Institute for Advanced Study mostly because of her work on cosmology. The institute, founded by New Jersey department-store magnate Louis Bamberger, is completely independent of Princeton University, which lies a mile or so to the east, across a golf course. The scientists at both institutions work closely together, however. Albert Einstein, for example, the institute's first hire, would frequently shuffle over to the university to give or attend talks and consult with colleagues. Combined, the university and the institute allow Princeton to rival Cambridge or Pasadena or the other Cambridge as one of the world's leading centers of cosmology—which was tantamount, when Seager arrived, to astrophysics. When Geoff Marcy's partner Paul Butler had come through in 1996 to give an early talk about exoplanets, the local astronomers listened politely and asked a couple of questions, but they didn't quite understand what the fuss was about.

The fact that Seager had moved on from cosmology didn't bother John Bahcall, though. Bahcall was head of the institute's School of Natural Sciences, and the man who hired Seager. He was a cosmologist like everyone else, an expert on, among many other things, neutrinos, a type of elementary particle so ephemeral that it could speed through a wall of lead a trillion miles thick without noticing there was anything in its way. Bahcall was also a key figure in getting NASA (and by extension, Congress) to build and launch the Hubble Space Telescope. Bahcall's policy was to find the best people he could and bring them to the institute, whether they were in the mainstream or not. The institute's other natural science faculty members were less certain, however. "I learned after he died [in 2005] that they weren't so sure about bringing me on," said Seager. "They knew I'd be working on exoplanets, but who would I work with?"

The answer came about a year later. Although it had taken them a little longer than their counterparts at Harvard, several of the Princeton cosmologists had begun to take an interest in exoplanets. One was David Spergel, a MacArthur "genius" grant winner for his work on cosmic structure and other esoteric ideas. In the early 2000s, Spergel was still deeply involved in a satellite called the Wilkinson Microwave Anisotropy Probe (WMAP), which would dramatically improve on the data beamed down from the Cosmic Background Explorer a decade earlier. Another was Edwin Turner, whose specialty was gravitational lenses—a phenomenon, predicted by Einstein in the

1930s and finally found in the 1970s, in which a massive foreground object, such as a galaxy, warps and distorts the light from a more distant background object.

Turner, Spergel, an aerospace engineer named Jeremy Kasdin, and a few other interested parties, including a couple of students, began meeting informally once a week to talk about exoplanets in general. Among other topics, they also talked about one of the projects NASA had been dreaming about for twenty years, at least, but which was now, thanks to the discovery of actual exoplanets, getting some research funding. This was the Terrestrial Planet Finder (TPF), a space telescope powerful enough to image exoplanets directly, and maybe even search their atmospheres for telltale signs of life. "So I was sort of the local resident expert on exoplanets," said Seager, "and David Spergel said, 'Do you want to join this team?' And what followed was one of the most enlightening and invigorating series of discussions—I won't say that I've ever had, because this happens a lot in exoplanets, but it was amazing. The word *no* was not really allowed. It was like, 'Can we detect lightning on an exoplanet?' That kind of thing. It didn't get documented, because we weren't fast enough to get it on paper. And Spergel came up with the pupil idea that summer."

The pupil idea is something people at Princeton still shake their heads over. The biggest problem with imaging a planet directly is that planets and their stars are very close together, and stars are vastly brighter. It's like trying to pick out a candle sitting next to a searchlight. So you need to blot out the star's image somehow, and the simplest answer—just paint a black dot on the telescope, or the equivalent—is really hard to do for complicated reasons involving the laws of optics. David Spergel became fascinated with the problem, but he didn't know much about optics. So he took home a textbook one weekend, read it through, and showed up the following week with a radical proposal. Instead of blotting out the starlight directly, you could achieve the same effect by putting a mask over the telescope's opening with a cutout in a shape resembling a cat's eye.

Because of the way light diffracts, or bends, at the edge of the cutout, the opening redirects starlight so that it more or less disappears in big areas of the image. It's a bit like the way ocean waves bend around promontories and other features on a shoreline to leave some areas calm and others choppy. If there's a planet in the place where the starlight has been removed, you can see the planet, faint though it is. "This is a completely new idea," Michael Littman, an optical engineer at Princeton and a member of the discussion group told me at the time. "But once you see it, it's obvious. I kick myself that I didn't see it myself." Ed Turner was a little more colorful: "My jaw dropped when I heard about it," he said.

Seager found the conversations exhilarating, and they led to collaborations with Turner and a Princeton grad student named Eric Ford on how TPF might detect the presence of life on an Earth-like exoplanet if and when the planet were found—a difficult job, given that all you'd see would be a single dot of light, but they thought of a couple ways to do it. But TPF wasn't going to happen for years, and Seager had plenty of other research ideas. She thought she'd given up on searching for transits herself, but then she started up another search, working with a postdoc named Gabriela Mallén-Ornelas. Mallén-Ornelas was from Chile, where several of the world's biggest, most powerful telescopes are located, in the high desert in the north of the country. In return for providing the sites, Chile gets 15 percent of the time on the telescopes. "She and I sat down and said, 'What can we do with all that telescope time?'" They decided to search for transits. "We did it all by ourselves, starting from scratch," Seager said. "Unfortunately," she continued, "we failed in the end because if you do a complicated experiment you need to get every single step right, and we made a small mistake. We didn't appreciate how hard it was going to be to work with very faint stars."

In retrospect, Seager thinks John Bahcall suspected she might fail. "He did say to me afterward, 'Now you really understand data. That's maybe all I wanted you to get out of this project.'" What Bahcall was getting at was the fact that Seager was a theorist and a builder of computer models. She had spent most of her career in the realm of the abstract, but it was important to go up against the real world of observations as well. Bahcall also thought it was important to take scientific risks. "At that time," she said, "I was traveling all around the country giving talks. There was always one person in the audience who would come up to me afterward, usually some middle-aged man, some professor, who would say, 'This is so exciting, if I was a young man I would be doing this.' I don't know if they would though. It was a big risk to invest so much in a field that a lot of people thought would go nowhere. But

I had an advantage. At the institute, I was supported to do anything I wanted, and I had John's guidance. He didn't say, 'Do this, do that, I advise this, I advise that.' I'd say, 'I am so excited about this paper I'm working on, because it's going to be so useful for observers.' And he would say, 'It's not going to be useful if you don't get it published.' " Bahcall mentored her, she said, "in this really subtle but persuasive way."

For plenty of young scientists, Bahcall could also be a little bit terrifying. Once a week, for many years, he would host what was known by astrophysicists around the world as Tuesday Lunch, in a dining room off the main institute cafeteria. Physics and astrophysics faculty from the institute and the university, along with postdocs, grad students, and eminent visiting scientists, would sit around a long, U-shaped table, with Bahcall at the head. First he'd call on the person sitting next to him, who was always the visiting scientist who had just finished giving a formal talk in the hour before lunch. Bahcall would ask him or her to speak briefly on an entirely different topic, off the cuff. Then he'd go around the room, sometimes in order, sometimes at random, asking whomever his eye landed on to speak for a few minutes about his or her research project. He was always perfectly friendly about it, but if he asked a follow-up question that exposed a lack of rigor or knowledge on the speaker's part, it was considered deeply embarrassing.

By Seager's lights, though, he was simply a great guy. "People told me a story about a time when child care at the institute was a real problem—mostly for the wives, because the postdocs were mostly male." There was a day care center on campus, but they wouldn't take infants, and they didn't seem interested in changing that policy. "One day the wives spoke to John," she said, "and the next day they took infants. I mean John had real power and authority. And his idea was just to do whatever could help the postdocs become great scientists."

With a Ph.D. from Harvard, an appointment to the Institute for Advanced Study, and John Bahcall's blessing, you'd think Seager would have had plenty of job offers. The transit project didn't work out, it's true, but shortly after she'd arrived, she and her Ph.D. adviser, Dimitar Sasselov, had written an important paper together. As a planet transited in front of a star, they realized, some of the starlight would shine through the planet's atmosphere. Whatever elements or compounds were present in the atmosphere would absorb the starlight at specific wavelengths—specific colors of light. They'd create the same kind of spectral lines that Mayor and Marcy were using, except that these absorption lines were coming from the planet's atmosphere, not the star's.

When the planet was in front of the star, those extra lines would be superimposed on the star's own lines; when it went behind the star, the extra lines should disappear. Seager and Sasselov figured out which of the lines should be most easily visible in a telescope; they should come, the co-authors said, from sodium compounds. A couple of years later, Dave Charbonneau and Tim Brown used the Hubble Space Telescope to look for this effect in their original transiting planet, HD 209458 b. Sure enough, there they were. It was the first detection ever of an exoplanet's atmosphere. "As a result," Charbonneau told me, "Sara is really credited with a lot of those first calculations—because they turned out to be right."

Still, when Seager went out to talk to potential employers, she got mixed feedback. " 'This will never happen,' " she would hear, "this" being transits. "That's what a lot of people said. When there was one transiting planet, and then an atmosphere detected through this method I had laid out, they'd say, 'Oh, it's just one object, one method, one success.' And I said, 'Look, we're going to have so many transits we won't know what to do with them.' " But since she was interested in planets many of the people she was talking to were in departments of Earth and planetary sciences. "They're used to holding things in their hands. Like drilling cores and things like that. And they just couldn't believe that it would ever happen." Nowadays, she said, "you get exoplanet papers that are just . . . well, I hesitate to use the word *garbage*, but they're not always backed up with serious work. The attitude now is, 'That's okay, if it's exoplanets it gets the stamp of approval.' "

When Seager finally got a job offer at the Carnegie Institution of Washington, in Washington, D.C., some people strongly advised her against going, in part because she still had two years to go at the Institute for Advanced Study. "I asked John, 'What should I do?' And he might have said, 'You don't have any other offers, of course you should go.' But he thought about it for a while, and he just said, 'Vera did all right.' " Bahcall was talking about Vera Rubin, an observational astronomer who began working in the late 1950s, when there were very few women in the field. Rubin was one of the first to show that galaxies rotated faster than they had any business doing, suggesting that the gravity of some unseen matter was influencing their rotation. Many of her colleagues refused to believe her for years. Some even implied she was incompetent. But Rubin's observations eventually forced astronomers to accept the existence of dark matter, a mysterious, invisible, so-far-undetected material that outweighs ordinary stars by a factor of ten to one. It's gravity from the dark matter surrounding them that makes visible galaxies spin so fast. Today, Rubin is widely recognized as one of the most influential astrophysicists of her generation. Vera did indeed do all right.

about the other one. "Natalie is awesome. She is incredible!" Fischer exclaimed once when I mentioned her former office mate's name. "I love Debra!" Batalha said in a similar situation. I've never heard male astronomers talk quite like this. They might praise others' brilliance or technical skill or professional achievements, often with enthusiasm, but they rarely talk about what wonderful people their colleagues are. It would be an unfair stereotype to suggest that women in astronomy are warm and nurturing while men are all about nuts and bolts and pissing matches. But after nearly three decades of writing about astronomy, it's pretty clear to me that the entry of women into astrophysics in large numbers over that time has improved the overall empathy quotient in the field.

If they were unlikely to end up as grad students in astrophysics, Batalha and Fischer were even less likely to end up with such high profiles in the scientific community—Batalha, as Bill Borucki's deputy on the Kepler Mission and a frequent public spokesperson on the project, and Fischer, as a full professor of astronomy at Yale, where she was recruited a couple of years ago to bring some prestige to a department that had long been considered the poor relation of Harvard and Princeton. It's not that either woman's work in grad school was anything but outstanding; it's just that only a few graduate students of either gender, even at a top-rated program like the one at Santa Cruz, go on to be so prominent.

For Batalha, the first step into science came in her freshman year at Berkeley. "I didn't know from a young age that I wanted to study science at all," she told me one afternoon in her office, right next door to Bill Borucki's at NASA's Ames

Chapter 7

INVASION OF THE FEMALE EXOPLANETEERS

A T ABOUT THE same time Dave Charbonneau and Sara Seager were blazing their way through Harvard's graduate program in astrophysics, two women were moving more quietly through a similar program on the other side of the country, at the University of California, Santa Cruz. Unlike their Canadian counterparts, neither Debra Fischer nor Natalie Batalha had started out studying cosmology. Neither had even been a hard-core physics geek in college—or in Batalha's case, at least, not when she started out. Her father worked in construction and her mother did accounting. They were convinced that their daughter would have the best shot at a good life if she majored in business, and she'd arrived at UC Berkeley in the late 1980s assuming that she'd do just that. Fischer was an even more unlikely astrophysicist: She'd gotten her undergraduate degree in nursing at the University of Iowa, and worked for several years afterward as a scrub nurse in open-heart surgery and in intensive care units.

Neither Batalha nor Fischer can say enough good things

Research Center. "I wasn't the kid who read science fiction, I wasn't the kid who watched Carl Sagan's *Cosmos*, I wasn't any of that. And yet I grew up with an innate sense of wonder, I guess is the best way to describe it. And perhaps curiosity, or maybe an eye for the beautiful in nature. I guess that makes it sound a little bit simplistic." Batalha did find the idea of human space exploration to be romantic. "Being an astronaut—of course, for every kid growing up at that time, it seemed like the most glamorous job ever. I mean all of us thought that."

It wasn't until she took her first college physics course, though, that things began clicking into place. "Somehow," she said, "the physics I learned in high school didn't inspire me. I think it's because it lacked the connection between mathematics and science that is so fundamental—being able to write down an expression that predicts what will happen in the future, and just putting the universe in an orderly context. I'm not a math buff by any means, but the beauty of it is what struck me."

This revelation didn't turn Batalha into a scientist, but an internship she did the following summer at the Wyoming Infrared Observatory, near Laramie, arguably did. It was a pretty small-scale operation, which meant that a summer intern had a chance to do some actual research. Her supervisor was an astronomer named Gary Grasdalen, "a very brilliant man," said Batalha. "He passed away quite young, shortly after I left." Grasdalen gave her a problem to solve that had been stumping everyone. "It was just kind of a little technicality thing," she said, "but I was able to solve it. And that surprised me. And it was fun. And the science that we got out of it as a result was

Natalie Batalha (Courtesy of NASA)

very gratifying." She decided to stick with astrophysics. "Gary told me that when I got back to Berkeley I should go knock on Gibor Basri's door."

At the time, Basri hadn't yet begun working directly on exoplanets, although he eventually would; his research concentrated on stars and on brown dwarfs—the objects Michel Mayor was originally interested in as well. But then, this was in the late eighties, when almost nobody was working on exoplanets. "Geoff Marcy was actually kind of there at the time as well," said Batalha. "He was a professor at San Francisco State, but he was going back and forth." Everyone knew Marcy was searching for planets, but, said Batalha, "there was this kind of aura at Berkeley at that time that it was never going to go anywhere. Nobody took it seriously."

Guided by Basri, Batalha majored in astrophysics, doing her undergraduate research on young, Sun-like stars. She then went on to Santa Cruz. Her thesis adviser there was Steve Vogt, who built the Hamilton Spectrograph at Lick Observatory, the instrument Geoff Marcy and Paul Butler would use to confirm Michel Mayor's discovery of 51 Peg b, and then use to discover their own planets. When the spectrograph was built, it had to go through a commissioning phase, running through a series of test observations to make sure it worked properly. Once it was fully operational, the only way to get access to the device was to apply for telescope time at Lick, but Vogt gave Batalha some of the data from the commissioning run to use in her thesis. "I ended up producing something," she said, "and was invited to give a talk at a conference in Vienna."

A week after the talk, there was another conference in Florence, Italy. Since she was already in Europe, and since it was relevant to her own research, Batalha decided to stay on. It turned out to be the meeting where Michel Mayor would announce the discovery of 51 Pegasi b. The title of the Florence gathering was "Cool Stars, Stellar Systems and the Sun"—and in fact, said Batalha, many of the people working in exoplanets today got their start studying stars. "The reason for that is quite simple," she said. "The way we find exoplanets is by observing changes in starlight." This is true whether you're searching for radial-velocity wobbles or transits. "So you have to be a stellar spectroscopist, you have to be a stellar photometrist, you have to understand stellar activity and how it manifests itself on the surface of the stars. You have to have a deep understanding of all of it in order to pull out the exoplanet signal."

As for Mayor's announcement itself, she recalled, she didn't realize at the time how important it was. Neither, as far as she could tell, did others in the room. "It was kind of nondescript," she recalled. "We didn't really know what was happening, except there was a camera crew that came in to film it. So I knew it was something kind of important, but I didn't really pay too much attention. And Michel Mayor got up there and gave the talk, and I guess as a young student I didn't appreciate it so much. But there really wasn't a sense of excitement in the room. It just kind of passed. And I am sure there were people sitting there who did appreciate it, but there was no buzz. It didn't really sink in for people until later, the import of it."

Batalha ended up taking a while to get through graduate school. She'd married one of Gibor Basri's postdocs, a Brazilian astronomer named Celso Batalha. "He was funded by the Brazilian government," she said, "so he had to go back after his postdoc was over. All during this period, I was moving back and forth between Brazil and the U.S., taking leaves of absence. I actually spent my last year down in Brazil and wrote my thesis while I was there." She stayed on in Brazil to do her own postdoc after finishing her Ph.D., still working on the astrophysics of stars. While she was in Brazil, she got an e-mail from her old office mate Debra Fischer. "Debra said, 'You know, there's this guy at Ames who's building a telescope at Lick to do transit photometry, and they really need help.'"

"This guy" was Bill Borucki. By this time, he'd been working actively on his planet-detection satellite for more than a decade. The first formal proposal had gone up to NASA headquarters in 1992 as part of the Discovery program. Discovery had been designed to accommodate relatively inexpensive missions—in the hundreds of millions of dollars rather than the billions—that would mostly focus on the solar system. According to Borucki, the thinking at NASA headquarters was, "We'll let people who are interested in other solar systems propose too." So Borucki sent a proposal up in 1992. Back then, Kepler was known by the much klunkier name FRESIP, for FRequency of Earth-Size Inner Planets. NASA's response, in essence: It sounds like a great mission, but the detectors you'd need to pull it off don't exist.

So Borucki and his collaborators went back and built the detectors, and came back in 1994 with another proposal. "They looked at us," Borucki told me, "and said, 'This is a space telescope, like Hubble. Hubble cost several billion dollars. You think you can build this for three hundred million dollars?' They just threw the whole proposal out." So he came back again in 1996. "We priced it three different ways, and showed how we could do it on the available budget. They said, 'Okay, your price is plausible, but nobody has ever done photometry on thousands of stars at once, so your proposal is rejected. Go out and build an observatory and prove it can be done.'" By now, Batalha's comment about Borucki starts to make sense: "Bill has this personality trait where negativity just rolls off of him. He doesn't accept it at all."

So he and his crew went up to Lick Observatory on Mt. Hamilton, outside San Jose, where Geoff Marcy and Paul Butler had spent endless nights searching for wobbling stars. They took over a small, unused observatory building, fixed the place up, and put a telescope inside. Having been rejected so many times already, Borucki had a hard time convincing anyone at NASA to give him funding. "We had a charge card," he said. "Everything had to be bought on a charge card. We'd buy this part, and that part— I shouldn't say that. You can't do that. It's illegal." He was clearly not afraid, at this point, of getting in trouble.

"It was a great little telescope," he said. "Visitors would come by and I'd tell them, 'Here's the biggest telescope on the mountain.'" They'd look at him blankly, since it had only a four-inch light-gathering mirror—so small that even a half-serious amateur astronomer would sneer at it. The Shane telescope, a short distance away, had a mirror 120 inches across. But while Borucki's telescope couldn't gather a lot of light, its field of view—the amount of sky it could see all at once—was bigger than the Shane's. When you're trying to peer deep into the universe, a narrow field of view is what you want; it lets you get the most possible light out of just a few, very distant, faint objects. Borucki was less interested in the amount of light he could gather. He needed to measure variations in light, not total amounts, and he needed to do it for ten thousand stars at once. "So we did this," he said. "We showed it could be done."

The telescope was named Vulcan. "No"—Batalha would say a decade later, seeing my eyebrows go up—"it had nothing to do with *Star Trek*." Vulcan was a hypothetical planet orbiting between Mercury and the Sun. The French mathematician Urbain Le Verrier proposed its existence in 1859. This wasn't the first time Le Verrier made such a prediction. In 1846, he suggested that irregularities in the orbit of Uranus could be explained by the gravity of an unseen planet. He told observers where to look for it. When they looked, there was Neptune. Mercury also had an odd orbit, so Le Verrier invented Vulcan, named for the Roman god of fire. Things didn't work out so well this time. Despite extensive searches, Vulcan was never found. It turns out that the irregularities in Mercury's orbit come from a different source. The tiny planet is so close to the powerful gravitational field of the Sun that Newton's version of gravity, which is normally sufficient to describe planetary orbits, doesn't work. You need Einstein's general relativity to explain it, and that theory wouldn't be published until 1916.

Still, it was a good name for a planet-hunting telescope. When Debra Fischer first told Natalie Batalha about the Vulcan telescope, Batalha noted it with no more than mild interest. But over the next year or so, she said, "it cogitated in my brain," and, after about a year of cogitation, she e-mailed Bill Borucki. Her special area of expertise in stellar astrophysics was starspots—sunspots on stars. "I said, 'You know, I'm really interested in how you can disentangle a transit signature from spot signature." It's a good question: If you see a tiny, repeating, regular dip in starlight, how do you know it's a planet moving in front of the star and not a dark starspot on the surface, rotating along with the star? "As luck would have it," she told me, "or maybe *luck* isn't quite the right word, that was one of the weaknesses in an earlier Kepler proposal. Ed Weiler at NASA headquarters, whose specialty is magnetic activity in stars, had expressed doubt about Kepler's ability to disentangle these two signals, and so I proposed to Bill that I come think about this problem. And he invited me to come, and I came in February of 2000."

In part, she came because she was following the money. NASA was getting more and more interested in exoplanet searches during the late 1990s; it was clearly something the public was interested in, and the public ultimately paid the bills. The agency had honed its public relations machinery during the 1960s, when it successfully painted the race to land men on the Moon as a grand adventure for humanity and a quest for scientific knowledge. In the 1990s, NASA's PR apparatus was geared up once again, this time to tout the International Space Station. Just as with the Moon landings, this project was a geopolitical maneuver. Its real function was to keep Russian rocketeers usefully employed after the downfall of the Soviet Union, so they wouldn't be tempted to sell their expertise to Iran and other dangerous clients. What the public mostly heard, however, was that it was a major step forward to living and working in space, and an orbiting platform for amazing science.

In fact, the space station has turned out to be a lousy platform for science, as scientists had said all along that it would be. But the agency did do spectacular science, with the Hubble

Space Telescope, the Spitzer Space Telescope, and other orbiting observatories; with its probes to Mars, Jupiter, Saturn, and other planets; and with satellites such as COBE and WMAP, which tuned in to the ancient light from the Big Bang. As Bill Borucki and Geoff Marcy knew all too well, the agency had also been interested in planet-hunting for decades. It had even funded a number of studies, but had always concluded that a search for extrasolar worlds would have to wait for advances in technology.

In 1995 and 1996, however, Michel Mayor and Geoff Marcy had shown that we didn't have to wait after all. NASA officials quickly realized that the American public—their ultimate bosses—were captivated by the search for planets and for life. So the agency scrambled to repackage several of its existing projects, including its planned "Next Generation Space Telescope," into a program titled Origins. In January 1996, at the same meeting where Geoff Marcy stood up to announce his first two exoplanets, NASA administrator Daniel Goldin gave a talk about the Origins idea. It wasn't just a space-based program, he said. Origins would also include research on the origin of life on Earth, for example, and on the range of environments in which life could conceivably survive. And it would include ground-based searches for exoplanets. When Geoff Marcy went to the Keck II telescope a month or two after the conference to look for planet wobbles with the world's most powerful telescope, he used telescope time NASA had purchased from the Keck Observatory.

NASA's newfound commitment to funding planet searches and related research had also made a big difference to Bill

Borucki's Kepler project. "We'd been pushing this mission before people had found planets," he told me. "When they found planets, that helped our proposal." The Vulcan telescope on Mt. Hamilton had proven that Borucki and his collaborators could successfully monitor the brightness of tens of thousands of stars at once, and he'd gone back to NASA again in 1998 to propose Kepler. This was now the third time. "They said, 'That's just great that you can do that,'" Borucki recalled, "'but nobody has ever built a system that could maintain the required precision in orbit.'" What they meant was that the spacecraft has to rotate in order to keep its gaze focused precisely on its target stars. You do that with reaction wheels: The wheels turn in one direction and the spacecraft responds by turning in the other direction. But, he said, "reaction wheels aren't perfectly round, okay? Plato is the only one who believed in perfect circles, and he died a long time ago. Ours aren't Platonic wheels, they're real, and they shake, and so the satellite will shake, and the stars are moving across your pixels."

"So, of course, they rejected the proposal," he said. Again. NASA told him to go into the lab and build a test facility that proved it could be done. This time, at least, Borucki didn't need to use his credit card to build the test facility. "They gave us five hundred thousand dollars. That's nowhere near enough money, but Ames lent us another five hundred thousand dollars. Which was sort of scary because you have to pay five hundred thousand back. But it was progress." That funding made it possible for Natalie Batalha to join the project. But it wasn't just about following the money for her. "It's more than that," she said. "For me, this is such a profound quest. It's exploration in the very fundamental source of the word, right? Human beings have that seed in them to always search for new horizons—humanity in general, not just scientists. I think about that a lot. It drives me and it motivates me, and it makes me particularly interested in this area. And for me just as a career in general, I couldn't do anything that didn't have some kind of profounder meaning. Perhaps scientists all have that in common as well. I don't know."

Natalie Batalha decided to become an astrophysicist after she took a freshman physics class and spent a summer at an observatory in Wyoming. For her old office mate Debra Fischer, it didn't happen until she was in her late twenties; astrophysics was her second career. "I don't like to tell anyone about the earlier part of my life," she told me in her office at Yale. "It's just—I don't know why, because now it shouldn't matter. I'm tenured at Yale, what difference does it make? But I got my first degree in nursing, it turns out." This may not sound like such a terrible thing, but it's possible, although at this point in her career highly unlikely, that some of her colleagues would take it as showing a lack of seriousness. Even Carl Sagan, an astrophysicist and a professor at Cornell, was looked at suspiciously, and even denied admission to the National Academy of Sciences, because he also wrote popular books, appeared frequently on *The Tonight Show*, and hosted the *Cosmos* series on PBS.

After Fischer got a degree in nursing at Iowa, she wound up at the hospital at Case Western Reserve University in Cleveland, Ohio. "The whole time," she said, "I was much more fascinated with the instruments and the way things worked, the defibrillators and everything, and less focused on the poor sick people, actually." Even so, she was so immersed in the world of health care that when she decided to take her next career step, the plan was to go to medical school. Her boy-friend at the time—now her husband—was a medical student at Case; when he went out to San Diego for his residency in internal medicine, Fischer enrolled at San Diego State University to fulfill her premed requirements. "It was kind of a luxury," she said, "to go back to school and know that I could sort of play around a little bit. I had time to take a course on the history of jazz or classical music or art history." She'd always loved mathematics, so she also took some math and physics and, just for fun, an astronomy class. "I realized that astronomy was this amazing study of everything," she said, "with the insignificant exception of Earth. If you look at Earth compared to the universe it's nothing. It seemed so exciting."

By the time her husband moved on to a second residency, in cardiology, in San Francisco, Fischer had given up on medical school. She enrolled at San Francisco State University to do a master's in physics. "They had just hired Geoff Marcy—I think it was 1984 or something like that. I went observing with him at Lick," she said, "and just fell in love with the whole observatory and the process of looking out into the universe. It was a great opportunity." The obvious next step was to apply to a Ph.D. program, but Fischer wasn't so sure. "I thought, 'I'm too old, I'm already thirty, I shouldn't go back to school.'" So she taught some undergraduate physics courses at San Francisco State for a couple of years, but, she said, "I couldn't stop thinking about going back." She applied to several schools without a huge amount of confidence. "I remember Geoff saying, 'It's not up to you whether they accept you. It's out of your hands. If they decide you're too old, that's the way it is.'" Of the five schools she applied to, she got into four. She picked Santa Cruz, partly because it was the closest to San Francisco, where her husband was now a cardiologist, and where they now had a child. "Actually," she said, "two children, by the time I started."

Debra Fischer (Tony Rinaldo)

Like Natalie Batalha, Fischer worked on stars, not exoplanets. There simply weren't enough groups working on the topic yet to provide grad students with good research projects, and money hadn't yet poured into the field. Fischer's thesis was on using the element lithium as a marker for stars' ages. But at one point in 1997, she went to an International Astronomical Union symposium in Boston. Geoff Marcy had been there too. "It was a really exciting time," she said, "because Geoff was talking about the first three exoplanets, which was all they had at the time." She was sitting in economy, in a middle seat at the very back of the plane, and suddenly Geoff Marcy was standing there. "He'd come back from first class [so clearly *some* money was flowing], and he said, 'I've been thinking. Paul Butler has this great opportunity to go start an exoplanet project at the Anglo-Australian Telescope, so we'll need another person on my team. What do you think?'" As best as Fischer can recall, her answer was, "Are you kidding? This is so exciting! I will put my whole heart into this project."

And she did, driving up to Lick late in the afternoon some eighty days a year, taking data all night, then driving back down for her day job, first as a postdoc and then as a research astronomer at Berkeley. "It would be absolutely insane," she said, "but back then, every planet was a big deal." It was an even bigger deal when the team found signs of a second planet around the star Upsilon Andromedae, where they had found one already. At this point, no star had yet been shown to have a second planet.

Fischer's job was to take measurements of the star's radial

velocity over many different nights and try to see if they fit a curve. This was how Mayor and Marcy and Latham had been doing it for years. You can't simply aim your telescope on a star and watch it move; you take a reading every so often and plot it on a graph. Today, the star is moving toward you at such and such a speed. Another day it's moving away. Another day it isn't doing much of anything. (This is an oversimplification: Stars are *always* drifting toward or away from Earth as they bobble along in their independent orbits around the core of the Milky Way. The motions Mayor and the others were looking for were on top of that constant, steady drift.)

Over time, those measurements should trace out a curve representing a repeating forward and backward motion—the signature of a planet. If you expect the curve to move leisurely up and down on your graph over years, you wouldn't bother making a measurement every day. That's how Marcy and Butler missed 51 Peg; a planet with a four-day orbit isn't going to show any sort of obvious pattern if you look at it randomly once every few weeks or months.

If there are two planets orbiting the star, it gets more complicated. There are two radial-velocity curves now, with different strengths and periods, but they're superimposed. Each measurement gives you the combined effects of both planets' gravity on the star. Fischer appreciated this, but even so, she said, "it was horrible. It wasn't coming out right at all." Finally, she took the two curves that best fit the data, bad as they were, and subtracted them from the overall signal. When she did that, she recalled, "I saw the data doing this unbelievable extra sine curve. It sent chills down my spine. It looked like there was a *third* planet in there, and no one had expected it. I remember holding the plot," she said, "and walking across the campus to Geoff's office, thinking, 'Look at this, there's a planet with a 4-day orbit, there's one with a 240-day orbit, and there's one with a 2.5-year orbit, and they are *big.*'" It was hard to imagine how such a system could be stable, with three huge, tightly packed planets tugging not just on their star but on one another. "It has completely changed our vision of how planets form, how much space they really need, that sort of thing."

At just about that time, Bob Noyes called up from Harvard to say he'd been looking at the same star, and thought he had enough data to confirm two planets. "I knew it wasn't two planets, I knew it had to be three," said Fischer, and she was really distressed when Marcy proposed that the two teams combine their data and publish a joint paper. Once they saw her analysis, she realized, the Harvard team would see the third planet too, and get some of the credit. "I look back," she said, "and it's silly to have been so disappointed. But I was." Officially, the Harvard and Berkeley teams did get equal credit for discovering the first three-planet solar system beyond the Sun (a fourth may have now been found). Unofficially, without authority, and surely against their wishes if I should be foolish enough to ask permission, I hereby award full credit to the Berkeley group.

At the time, Marcy and Butler had been able to hone their technique to the point where they could measure a star's motion to a precision of three meters per second. That was good enough to find Jupiter- and even Neptune-mass planets around other stars, but not good enough to find an Earth-mass planet, even in a very tight orbit. (Butler's group in Australia would remain part of the Berkeley team, and when Butler later moved to the Carnegie Institution of Washington, the collaboration would continue.) The astronomers weren't satisfied. They didn't know how much better they could do, but they would try. "I remember," said Fischer, "Steve Vogt, Paul Butler, Geoff Marcy, and I would always sign our e-mails 'OMPSD,' which stood for 'one meter per second or death.' Steve, especially, loved that. But it was comical. Because of course you could never get to one meter per second."

Chapter 8

KEPLER APPROVED

WHEN BILL BORUCKI got the million dollars that allowed him keep honing the Kepler concept, his job wasn't just to convince NASA decision makers that his spacecraft would be stable, but also that it could reliably detect minute changes in light intensity. "They told us, 'It's probably going to take you several years to build it, because nobody's ever done anything like it before,'" he told me. "So we went out and we bought up all of the invar we could find"—invar is an alloy of steel and nickel that expands and contracts very little when the temperature changes. The alloy, invented in 1896 by a Swiss metallurgist named Charles Édouard Guillaume, proved so useful for constructing high-precision scientific instruments that Guillaume won the Nobel Prize in 1920 as a result. Albert Einstein had to wait until 1921 for his.

Despite NASA's pessimistic projection, said Borucki, "we got the design together in a few months, and then we got all the machine shops in the Bay Area building the parts for us. Got this whole thing built in a year, and got it debugged six months later. Then we had to figure out ways to simulate transits and measure them with high precision. We didn't care about accuracy."

For anyone but a scientist or an engineer, this probably sounds nutty, but it turns out to make perfect sense. Kepler didn't need to measure how bright a given star is. The scientists didn't care if this star is exactly as bright as that other star, or brighter or dimmer. They could afford to be inaccurate about that kind of measurement. All the satellite had to do was measure, with extremely high precision, the *change* in brightness when a planet passed in front of the star. "It's very much like what Geoff Marcy does with radial velocities," said Borucki. Marcy didn't care about a star's overall motion—it could be speeding toward the Earth or speeding away or sitting there like a bump on a log. All he cared about was the *change* in motion caused by an orbiting planet. "Geoff's accuracy . . . well, I would hardly call it poor, but he considers it poor," said Borucki. "But his relative precision is one part in a hundred million."

So how hard would it be to achieve the kind of precision Borucki needed to find an Earth—a precision of between ten and one hundred parts per million, or a millionth to a ten-millionth of a percent? "Think about those holes in a metal plate that represent stars," he said. "Now I slide a piece of clear glass in front of one of them. How much will the decrease in light be? For regular glass, it's 4 percent. That's way, way above ten or a hundred parts per million. So we take the glass and add an antireflective coating." That makes less light bounce off the glass and more pass through. "Well, now it's, you know, 1 percent or 0.5 percent. You're still hundreds of times too high."

No good. So in order to test the sensitivity of their detectors, they had to think of an entirely new way to create minuscule changes in light. David Koch, Borucki's longtime collaborator, came up with one. "We drill laser holes in a steel plate," said Borucki, "and we take a very fine wire and run it across the hole. Then we run a little current through the wire; the wire heats up and expands and it blocks more light." They built it. It behaved perversely. "Of course," he said, "when we applied the current, we got *more* light coming through, not less. Now, how can that possibly be?" It turned out that if the wire wasn't absolutely straight to begin with, the expansion would make it bend. It was no longer covering the widest part of the round opening, so it was actually letting more light through. "So we had to go back and redrill the holes as squares."

There was plenty more of this sort of thing, but by the time Natalie Batalha joined the group in 2000, Borucki and Koch had finished their experiments. They had convinced themselves that they could make precise (albeit inaccurate) measurements of the tiny changes in starlight that would signal the presence of a faraway, invisible world. They were writing up yet another proposal for submission to NASA's Discovery program. But it wasn't enough simply to prove that their space telescope would work properly; they also had to convince NASA that planetary transits were observable in principle. A planet would pass directly in front of its star only if the orbit were precisely edge-on. Since planetary orbits could come in any orientation at all, the Kepler team had to show that enough of them would line up correctly, purely by chance, to allow the spacecraft to find them in enough numbers to justify spending hundreds of millions of dollars on the mission.

Borucki had promised NASA that Kepler would measure the dimming of starlight with a precision nobody had ever achieved. To do that, however, the team needed a crucial piece of information. Say you measure the light curve of a transiting planet, and it dims the star's light by 1 percent. That tells you that the planet's diameter is 1 percent as big as the star's. *But how do you know how big the star is in the first place?* Without knowing that information, you can't learn anything.

Fortunately, Kepler can get the answer, by using a technique called astroseismology, which borrows its name from plain old seismology, the study of earthquakes. When an earthquake goes off, the shock propagates outward, but also downward. The downward shock waves travel toward the center of the Earth, then bounce when they run into a change in density—where the semi-molten rocky mantle meets the iron core, for example. Around the world, seismic stations pick up those bouncing shock waves, and seismologists use them to deduce the inner structure of the planet.

The Sun isn't solid like the Earth; it's a huge ball of gas. It's so dense, however, that it acts something like a huge blob of incandescent Jell-O. Turbulence in its upper layers makes the entire blob vibrate, in many overlapping frequencies at once. Decades ago, solar astronomers began to study the vibrations in the Sun by looking for radial-velocity differences from one part of the Sun to another. These differences are caused by the rising and falling of its surface under the influence of the vibrations. The pattern of vibrations depends very sensitively on the Sun's inner structure—on the temperature and pressure of its inner layers, mostly—but it also depends on the Sun's size.

With Kepler, astronomers could start doing the same sort of analysis for stars, and since knowing a star's size is crucial to calculating the size of a transiting planet, this was part of the Kepler Mission right from the start. As the project progressed through the 2000s, though, it began to run over budget. "NASA told us, 'You've got to cut things,'" said Borucki. "One of the things that came up for cutting was astroseismology." So the Kepler team cut a deal with a coalition of European stellar astrophysicists. The Europeans would get access to the Kepler observations before they were released to the public, and the Kepler team would get the astroseismology readings it needed in return.

The team also had to give a good answer to another obvious problem. We know the Sun has sunspots—dark blotches caused by the Sun's magnetic fields. We know that Sun-like stars have them too; they're called starspots, unsurprisingly enough. So how would Kepler distinguish between a dark starspot on the surface cutting down on the brightness of a star and the dark silhouette of a planet passing in front if it? This was Batalha's first assignment on the Kepler Mission. "I tackled the problem," she said, "from a stellar populations perspective." As a stellar astrophysicist, she knew that young stars tend to rotate quickly and have a lot of spots. That's problematic in two ways: First, the spots can masquerade as planets. In older stars like the Sun, that's not such a problem, since sunspots get fewer and a star rotates more slowly as it ages. The Sun rotates about once a month at its equator; a planetary transit should last only a few hours. It's easy to tell the difference. So Batalha's first job was to figure out what percentage of the Sun-like stars Kepler would look at might be too young to be trusted. "It came out to be about 25 to 30 percent of the sample," she said. "We basically showed that there would be enough stars left over that you'd still be able to detect substantial numbers of Earth-size planets."

Batalha and the rest of the Kepler team also had to figure out how they would deal with an even more common source of confusion. It turns out that the majority of stars in the Milky Way are part of multiple-star systems—doubles, triples, even quadruples, orbiting one another as they move through the galaxy. Our own Sun, with no companion star at all, is a little bit of an oddball. Theorists have long believed that single stars like the Sun are the best places to look for planets; the complex, ever-changing gravity of binary or triple stars would, they argued, make planetary formation difficult, or even impossible. So Kepler would keep its electronic eyes trained on singles. But since double and triple stars are so common— doubles, especially—it wouldn't be at all surprising if they were lurking in the background or the foreground, on the same line of sight as a target star.

At least some of these multiple stars would be orbiting each other edge-on with respect to Earth, purely by chance, each star eclipsing the other in turn. (It's called a transit only if one object is much smaller than the other—or appears much smaller. When the Moon passes in front of the Sun, it's called an eclipse because from our perspective, the Moon is big enough to block the Sun's light entirely.) If one of these eclipsing binaries was in the field of view, you'd be seeing the light from three stars in total, which would dip down to two stars during an eclipse, then back up to three, over and over.

If the single, target star and the eclipsing binary are very close together in the sky, Kepler won't see them as separate stars. "Say a background star is one thousand times fainter than Kepler's target star," Geoff Marcy explained, "but then it winks out by 30 percent. Then voilà! If you think you're looking at just one target star, the total dimming is a few parts in ten thousand." That's the same dimming you'd expect from a small planet. One way around this would be to take a second look with a more powerful ground-based telescope equipped with an adaptive-optics system. This is a technology originally developed by the military for spy satellites looking down on Soviet missile sites and other strategic targets: It uses a flexible mirror that constantly changes its shape to cancel out the blurring caused by Earth's atmosphere. Adaptive optics was declassified in the 1990s, and it turns out to be just as effective when you're looking up at a star rather than down at a missile silo. The Hubble Space Telescope is much smaller and less powerful than many ground-based telescopes; its major advantage is that it doesn't have to deal with the blurry atmosphere. With adaptive-optics systems, which are now common, telescopes on the ground don't either. Up to a point, anyway; adaptive optics isn't perfect. But unless the target star and the eclipsing binary are practically right on top of each other, adaptive-optics telescopes can separate two objects that Kepler sees as one.

So that's one way to rule out a background eclipsing binary star. Another is to see if the center of brightness in the image shifts at all. If all the brightness is coming from a single star, its position won't seem to change when the star dims. If it's coming from the combined light of three or more stars, the center will shift subtly. Yet another is to look in a wavelength of light other than the visible—infrared, say. When a planet passes in front of a star, the star's color doesn't change much. It just gets dimmer. But when one star eclipses another, the overall color does change, since it would be very unusual for two stars in a binary to have precisely the same color. An infrared telescope, which already sees a different mix of colors than a visible-light telescope, will see a dimming similar to what Kepler sees if there's a transiting planet. If there isn't—if it's an eclipsing binary—the infrared telescope and Kepler will see very different types of dimming.

Yet another way to rule out eclipsing binaries is to do it statistically. Based on what astronomers actually know about how many eclipsing binaries there are in the Milky Way, you calculate how many are likely to be in Kepler's field of view. "Then," explained Borucki, "we say for this tiny area around this star, here's the probability that there could be an eclipsing binary that would imitate it. Generally, we try to keep the uncertainties much less than 1 percent."

Even when they had ruled out false positives to the best of their ability, however—if a star had dimmed three times on a repeating schedule (ruling out one-time glitches) and had passed all of the false-positive tests the Kepler team threw at it—it could still be something other than a planet. To paraphrase Donald Rumsfeld talking about the unexpected problems that turned up after the invasion of Iraq, you can rule out the known unknowns, but that still leaves you with unknown unknowns. One more test, however, could transform a "planet candidate" into a bona fide, rock-solid planet detection. If you could detect not just a periodic dimming but also the radial-velocity tug an orbiting object imposes on its star, you could be a lot more confident there was a planet there. It would be the mirror image of what had happened when Tim Brown and Dave Charbonneau on one hand and Greg Henry and Geoff Marcy on the other had found the very first transiting planet, HD 209458 b, and quieted the last of the planet skeptics.

Ideally, the Kepler team would confirm all their planets this way. In practice, it wasn't going to happen. For one thing, it takes a significant amount of telescope time to get a radial-velocity profile of even a single star, and there's only so much telescope time available. For another, most of the stars Kepler is watching are far away—deliberately, so the satellite can look at a lot of stars at once. But that means they're generally dim, so radial-velocity confirmations require time at a very big tele-scope, with maximum light-gathering power. That's in even shorter supply than time on a smaller telescope, like the Shane at Lick Observatory. "We did a calculation at one point," said Batalha, "to come up with the number of hours on a telescope like Keck [in Hawaii] you would need to confirm every one of those planets and it was astronomical, literally. It was ri-diculous, the number."

Finally, Kepler is ultimately looking for Earth-size planets in their stars' habitable zones. For a star like the Sun the habit-able zone is about where Earth orbits, ninety-three million miles out or so. "Right now," Batalha told me, "an Earth-size planet in a one-year period around a Sun-like star causes a radial velocity that's too small for us to detect. So that type of signal we will not be able to confirm." To the untrained ear, this suggests that Batalha and Borucki and the other charter members of the Kepler team knew from the beginning that the mission could never actually achieve its goal. It could never point to a star and say with certainty, "There's an Earth-size planet orbiting in the habitable zone of a Sun-like star."

But that was never actually Kepler's goal. The original name for the project, remember, was FRESIP, for FRequency of Earth-Size Inner Planets. Kepler was designed to find the *percentage* of stars with Earths in the habitable zone. For that, you didn't have to prove that any one star had such a planet. "We'll never be able to eliminate all of our false positives," explained Batalha. "So we have to pick and choose which planet candidates to follow up on. What we really want to do is understand our false-positive rate in a statistical sense. So we have picked a subsample of stars that we will just throw every trick in the bag at—less than one hundred, maybe between fifty and one hundred. Seventy, maybe. And we'll take what we learn from that exercise and apply it to the rest of the catalog in a statistical sense. If we say to the public, here are a hundred candidates, and our false-positive rate is 10 percent, you know ninety of them are right and ten of them are wrong. You just don't know which ten. That is Kepler's objective, to determine if the Earth-size planets are abundant, not to figure out which ones they are necessarily."

That may sound odd at first, but it actually makes sense. If your ultimate goal is to find life on other worlds, it's obviously best to search for planets around the nearest stars, since they're the easiest to observe. But if you make them too near, you might not have enough stars close by to give you a fair chance of finding anything. How far away "near" is depends on how common Earth-size planets are. The best way of figuring that out is to do a broad survey of a huge number of stars first, which is just what Kepler is trying to do. "That's always been the plan," Batalha told me. "Figure out the fraction of stars that harbor likely planets, then design a mission to find the ones closest to us. Do I have to look out fifty parsecs [about 160 light-years] or two hundred [about 640]?" That's what Kepler is meant to do. If it determines, say, that 30 percent of Sun-like stars have Earth-size planets in the habitable zone, you can get away with surveying a couple hundred of the nearest stars to look for life. If the percentage is only 1 percent, you need to cast your net a lot wider.

The people at NASA headquarters understood this rationale, and agreed with it wholeheartedly. They had right from the beginning. The reasons they kept shooting down Bill Borucki's proposals were largely practical ones: He hadn't figured out all of the technology yet, or they weren't convinced he was being realistic about cost. They weren't necessarily right, but they had the final say. Borucki had four chances to give up and move on to something else, and didn't. I suggested to him during one of our conversations that I could imagine that a longtime NASA employee might become philosophical about getting shot down so many times.

"Ah. Okay," he said. "You can imagine that. Good for you." He didn't say it in a mean way. He was just amused.

So what does it really feel like, I asked.

"Not like that at all, of course." Natalie Batalha had told me rejection just rolled off Borucki's back, but that might have

been an oversimplification. "There's certainly discouragement," he continued. "If we were going to go propose again,
that's a lot of work and we had failed a number of times. Do
you really want to be a consistent failure or just an occasional
failure? See, you think about things like that. One of our team
members was Carl Sagan and we would submit and fail, submit and fail, submit and fail, and then he came down with
cancer. I asked him, Did he really want to be a member of the
team when we proposed again? So, he wrote me a nice letter
that said, Yes, he felt the mission was an excellent mission, we
were going to accomplish great things and he was so enthusiastic about it. He wanted to participate and he was expecting
to feel better in the coming months and he would work harder
with us."

Sagan died a few months later. "I still have that letter," said
Borucki. "So, it's something that we have to go through. We
have to overcome our anger at people being so stupid as not to
immediately fund the mission. We have to work with them to
look at the suggestions from the review panel and ask, How
can we incorporate those? Where can we get the money to do
the new studies that are required? That means we go to headquarters, we go to Ames, people at different universities have
to find funds to come up with a new and better proposal. So,
yes, there's discouragement. There's even a little bit being
pissed—a little bit. But if you've got an idea that you think is
so valuable, so important to mankind's future, then you're
willing to accept that. You're willing to say, Okay, I put up
with all this crap, this discouragement, and I move ahead."

Not everyone reacts the same way. There was a Venus

mission that went head-to-head with Kepler the last time Borucki went into the competition, in 2000, and a Jupiter mission as well. "Everybody knew that the Venus mission was good," Borucki said. "Everybody knew that the Jupiter mission was superb. But when headquarters chose our mission, these people just stopped. They didn't propose on the next round. I think some of them are proposing again now but there were a couple of opportunities that they did not compete for. I don't know whether they changed teams or changed people or what's happened, but I can't blame them."

Writing a proposal for a mission costs several hundred thousand dollars, he said, and it's not a lot of fun. When you get downselected—in NASA jargon, that means you've made the final round—they send you a bunch of questions. "You have twenty-four hours to answer them, okay? But one of the questions that they sent us was itself a twenty-part question on the communication system. What coding would we use for x-band, a-band, things like that. What's the reception rate? What antennas were we going to use? Twenty parts to one question, twenty-four hours, that's typical. It's an enormous effort to write a successful proposal for these things."

Was there actual champagne involved when Kepler was finally selected? I asked.

"Yes, there was. We were all delighted. Where's my bottle? There's a glass up on the shelf, but the champagne bottle is here as well if someone hasn't run off with it. In any case, it was Moët."

Chapter 9

WAITING FOR LAUNCH

KEPLER WAS APPROVED in 2000, but it wouldn't launch until 2009. It was clear to everyone in the exoplanet business that the satellite would revolutionize the science of planet-hunting—eventually. It would if Kepler was actually completed, that is, and launched without blowing up, and reached its intended orbit, and worked properly when it got there. None of those things is ever guaranteed with space probes. Even if they were, it would have been preposterous for hundreds of scientists to sit idly by waiting for Kepler for the better part of a decade when there was so much they could do in the meantime to chip away at the question of how many stars have planets, and what sort of planets they are.

So while Bill Borucki, Natalie Batalha, and the rest of the Kepler team began the long, careful process of putting together a space mission, Geoff Marcy, Paul Butler, Steve Vogt, Debra Fischer, and their collaborators and postdocs and graduate students kept measuring radial velocities. Paul Butler took the job he'd been offered at the Anglo-Australian

Observatory, and then moved on to the Carnegie Institution
of Washington, but he'd remained part of the team. (Like all
the astronomers at the Carnegie, Butler joined the Department of Terrestrial Magnetism, whose delightfully archaic
name became obsolete in 1929, when the department completed its charter task of mapping the Earth's magnetic field.)

Michel Mayor and his group in Geneva kept at it as well,
taking radial-velocity measurements, trying to squeeze down
their uncertainties to find smaller and smaller planets. Sara
Seager kept theorizing about planetary atmospheres and about
what you'd be able to tell about an exoplanet—not just the
atmosphere, but also the surface, and even the vegetation—if
you could ever take a direct image if it, with something like
the Terrestrial Planet Finder. For a few years, she did it at the
Carnegie; then she moved on to take a faculty job at MIT.
Around the world, astronomers worked harder than ever to
add new worlds to the exoplanetary tote board. They wanted
to be able to say something meaningful about what was out
there, and they had no intention of waiting for Kepler if they
didn't have to.

And there were a lot more exoplaneteers now than there
had been just a few years earlier. (They weren't calling themselves that yet, although the term would eventually become
almost universal. Dave Charbonneau is pretty sure he was the
first to use it, sometime in the late 2000s.) When Marcy and
Mayor announced their first exoplanets, they were operating
way outside the mainstream of astronomical research. But by
the time Debra Fischer got into the planet business in the late
1990s, she told me, "it felt like there was a tidal wave com-

ing." The approval of Kepler might have been the crest of the wave, but the swells that came before it were considerable. It wasn't just a bigger effort from existing groups; funding requests for new searches were also starting to appear in grant-makers' in-boxes.

Many of these new proposals were for transit searches, inspired in part by the discovery of HD 209458 b, by Geoff Marcy and Greg Henry, neck and neck with Tim Brown and Dave Charbonneau. The tidal wave was also influenced by the community's growing awareness of the Kepler project, generated through Bill Borucki's quiet evangelism at conferences and in proposals through the nineties. One of the new searches, for example, was established by a young Harvard astrophysicist named Gaspar Bakos. Bakos had arrived from Hungary to take up a predoctoral fellowship in 2001. He brought along a project he'd begun working on as an undergraduate student in Budapest. It was and still is called the Hungarian-made Automated Telescope Network, or HAT-Net, and it was in essence a more elaborate version of the system Dave Charbonneau and Tim Brown had used to find the very first transiting planet, HD 209458 b.

Bakos wanted to create a network of small telescopes only a little over four inches in diameter—in the end he ended up using off-the-shelf telephoto camera lenses—that would automatically look for transits in a selection of stars. "I remember looking at his proposal as an external reviewer," Fischer told me, "and I said, Yeah, we should give this guy money, this is brilliant." They gave the guy money, and Bakos, working with Harvard's Bob Noyes—the same astronomer who inspired

Dave Charbonneau and Sara Seager and a generation of young astronomers to go looking for planets—and Dimitar Sasselov, Seager's thesis adviser, began putting the project together. Ultimately, it would have telescopes in Arizona, Hawaii, Israel, Australia, Namibia, and Chile.

Another automated survey, created by a half dozen universities in the United Kingdom and called the Wide Angle Search for Planets, or WASP, went online in the 2000s as well, with telescopes in the Canary Islands and South Africa. Dave Charbonneau, Tim Brown, and Ted Dunham followed up on their own dramatic first transit discovery by creating the Trans-Atlantic Exoplanet Survey, or TrES, with tiny telescopes in the Canary Islands, Arizona, and California. Yet another search grew out of Marcy and Butler's Berkeley-Carnegie collaboration. Known as the N2K Consortium, it set out to look for both radial velocities *and* transits in two thousand bright stars that hadn't yet been studied (N2K stands for "Next 2,000," as anyone who survived Y2K might have guessed). As HD 209458 b had made clear, you can begin to understand what a planet is made of only if you have both its mass and its size, so you can calculate the density.

The two thousand stars in the N2K survey aren't just bright; they're also known to be rich in "metals." This means something very different to astronomers than it does to anyone else. In astronomy jargon, a metal is any element other than hydrogen or helium, which is all of them. So while everyone agrees that iron and aluminum are metals, astronomers also include oxygen and carbon and nitrogen and neon in that category. Debra Fischer had been one of the first exoplaneteers to show that planets are more common orbiting metal-rich stars than metal-poor stars—not surprisingly, since planets are made in large part of these elements—so looking at these stars was a way to boost the odds of finding exoplanets. N2K uses giant telescopes in Hawaii and Chile to look for the radial velocities, and small, automated telescopes at the Fairborn Observatory in Arizona to look for transits.

And these were only a few of many transit searches that were under way. "When I attended a conference on extrasolar planets in Washington, D.C, in the summer of 2002," wrote University of Toronto exoplaneteer Ray Jayawardhana in his 2011 book *Strange New Worlds*, "transits were all the rage. Keith Horne from the University of St. Andrews counted two-dozen transit searches in the works, employing a variety of instruments ranging from wide-field cameras that used commercially available 200-millimeter Canon lenses to the majestic 4-meter telescope on Cerro Tololo, Chile." But at the time of that 2002 conference, he continued, almost three years after HD 209458 b had been confirmed, "no other discoveries had been reported."

That wouldn't change for another three years. Finally, in 2005, Greg Henry, under the N2K banner, picked up the transit of a planet called HD 149026 b, midway in size between Neptune and Jupiter. Just as in the case of HD 209458 b, the first transiting planet ever found, this one had originally been discovered by the radial-velocity technique, the planet-induced wobbling of its star, so that when Henry got the transit, the size and mass could be combined to figure out the planet's density. Surprisingly—or maybe unsurprisingly, since

the unexpected is routine in the world of planet hunters—
the new planet was denser than anyone had anticipated. HD
209458 b had been *less* dense than anyone had expected. HD
149026 b was bigger than Jupiter, but was structurally more
like Neptune, with a relatively large, solid core of maybe seventy Earth masses surrounded by a relatively smaller atmosphere.

Astronomers were finally beginning to get a handle on
what exoplanets were actually made of, and with all these surveys and more in operation, they'd keep adding to the list of
planets whose densities they could calculate. It was slow going, however. They had to look through Earth's murky atmosphere, and our planet's rotation kept them from being able to
stare continuously at any one star. It would be impossible, as
Batalha had said, to find a planet as small as the Earth orbiting
a Sun-like star.

It would be impossible, at least, with either transit searches
or radial-velocity searches. There was another way to look for
planets, however, based on a theoretical idea invented, though
never proposed as the basis for actual observation, by Albert
Einstein. When Einstein came up with his theory of general
relativity back in 1916, one of its implications was that space-
time would literally be warped by massive objects. The Sun,
for example, should cause enough of a distortion in space that
a ray of light passing by would be forced to change course. If
the Sun and a random star happened to be close together in
the sky, light from the star, speeding straight as an arrow for
most of its journey across the universe, would change direction slightly. The effect for observers on Earth would be that

the star seemed slightly out of position, compared to where it would be if the Sun weren't there. In 1919, astronomers took advantage of a total solar eclipse to try to measure the effect (there would be no point in trying at any other time, since you could never see nearby stars if the Sun were blazing away normally).

Sure enough, Einstein was right. LIGHTS ALL ASKEW IN THE HEAVENS, blared the headline in the *New York Times* on November 10, 1919. "Men of Science More or Less Agog Over Results of Eclipse Observations. EINSTEIN THEORY TRIUMPHS. Stars Not Where They Seemed or Were Calculated to Be, but Nobody Need Worry."

Decades, later, when he had moved to the Institute for Advanced Study, Einstein received a visit from a Czech electrical engineer named Rudi Mandl. Mandl had evidently become obsessed with the idea that under the right conditions, this light-bending effect could turn a star into a lens, magnifying the image of a more distant background star. He raised some money, traveled to New Jersey, and knocked on Einstein's door to beg the great man to confirm or deny this idea. Mandl was evidently relentless, possibly to the point of being somewhat annoying. So Einstein finally agreed, and it turned out that Mandl was right. Einstein wrote a short paper that appeared in *Science* in December 1936 that began: "Some time ago, R. W. Mandl paid me a visit and asked me to publish the results of a little calculation, which I had made at his request. This note complies with his wish." Einstein also wrote a private note to the editor of *Science*, thanking him "for your cooperation with the little publication, which Mister Mandl squeezed

out of me. It is of little value, but makes the poor guy happy."
In the paper itself, Einstein wrote that "there is no great chance
of observing this phenomenon."

As it turned out, Einstein was wrong. The phenomenon
now known as gravitational lensing was first observed in 1979.
Since then, this cosmic optical illusion has become one of the
most powerful tools in astrophysics. It's been used to measure
the size of the universe, for example, and to study galaxies so
faint they'd be invisible if they weren't enhanced by the magnifying effect of closer galaxies.

In 1992, a Polish-born scientist named Bohdan Pacyznski
came up with the idea of using gravitational lenses to figure
out what dark matter, the stuff Vera Rubin and others had detected but not identified, actually is. (Pacyznski, who died in
2007, also came up with the idea of HATNet—originally to
look for variable stars, subsequently hijacked to look for transits.
He was also one of the first to suspect that so-called gamma-ray
bursts are not feeble, nearby pops of energy but fantastically
powerful ones happening halfway across the universe.) One
possibility was that the dark matter consisted of subatomic matter, in the form of "weakly interacting massive particles," or
WIMPs. Another was that it was made up of brown dwarfs,
smaller than stars but bigger than planets. Theorists had begun
to call them massive compact halo objects, or MACHOs
("halo" meaning they surrounded the Milky Way). Whether
these names should be considered brilliantly clever or a bit too
cute is up to the reader.

Pacyznski realized that if MACHOs existed, they would
occasionally drift in front of more distant stars, briefly magni-

fying the stars' light through gravitational lensing. Paczynski called the phenomenon "microlensing" and he organized a survey called the Optical Gravitational Lensing Experiment (or OGLE—astronomers can't seem to resist), run primarily out of the University of Warsaw, to look for them. OGLE didn't find many MACHOs, but once exoplanets began showing up in the mid-1990s, it became clear that the technique would work for exoplanetology as well. If a star drifted in front of a more distant star, the latter would seem to spike in brightness for a short time, just as Einstein had said. But if the nearby star had a planet in tow, the spike might be followed (or preceded, depending on how things were arranged), by a dimmer spike as the planet drifted by in its turn.

If the planet were too close in to the star, the flares would come almost right on top of one another, so they'd be too hard to separate. So microlensing is especially good at finding planets—even planets as small as Earth—that are relatively far out. Transit and radial-velocity searches, by contrast, are best at finding close-in planets. That makes microlensing the best way to find Earths in the habitable zones of their stars, with one important caveat. A planetary system that randomly drifts in front of a distant star is only going to do it once. It won't come around again, so you can never do any follow-up observations. You'll never be able to measure a planet's physical size, or study its atmosphere. "It's not good for what I love," Dave Charbonneau told me, "which is characterizing the worlds I discover." But like Kepler, it could be an important statistical tool, telling astronomers how often what sorts of planets occur in what sorts of arrangements.

Only about ten planets have been found so far by micro-lensing, but that doesn't reflect poorly on its importance, says Charbonneau. "It's not fair to compare it to the huge efforts in radial velocities and the huge efforts in transits," he told me. "If we go ahead with a mission where there's a significant mi-crolensing element, it really could be transformative, because that would deliver thousands of planets that orbit far from their stars." Beyond that, he added, since the foreground star and planet move across your field of view, you're getting a snap-shot of the system rather than waiting until the planet com-pletes several orbits. "It's great if you're impatient, because it is the one method where you don't have to wait for even one orbital period to know if you're seeing a planet."

Finally, in the late 2000s, Charbonneau himself began fo-cusing his attention on a type of star that's not at all like the Sun. Astronomers group stars into categories based on their surface temperature and their intrinsic brightness (that is, how bright they'd seem if all stars were equally distant from Earth). The Sun is a G-type star, with a surface temperature of about 6,000° Celsius. For every star like ours, however, the Milky Way has about one hundred smaller, redder stars known as M-dwarfs, which are only around half as hot. M-dwarfs are so dim that they're very difficult to spot (astronomers haven't even found all of the M-dwarfs in our own nearby cosmic neighborhood). They're so much cooler than the Sun that their habitable zones—the orbits where water can remain liq-uid, and, in principle, nourish life—are much closer in. An Earth that orbited an M-dwarf once every 365 days would be frozen solid. An Earth orbiting close enough that its year was just a few days long would potentially be hospitable.

It turns out that both of these characteristics make M-dwarfs ideal places to search for small worlds. If you like looking for transits, the shadow of an Earth-size planet moving in front of a Sun-like star cuts out about 1/10,000 of the star's light. If an Earth transits an M-dwarf, it cuts out 1/1,000 of the light—a much bigger signal, and much easier to spot. If you're doing radial-velocity measurements, the fact that your Mirror Earth is so close, and also so much bigger compared with its star, means that it will yank the star around much more powerfully than we yank the Sun. And with both techniques, the planet's shorter year means you don't have to watch so long to see the orbit repeat itself. You can convince yourself it's really a planet much more quickly.

With all of these advantages in mind, Charbonneau put together the MEarth (pronounced *mirth*) Project, which would look for planets just a bit bigger than Earth, around two thousand M-dwarfs within the closest few tens of light-years. "We think," said Charbonneau when I asked him about it, "we have the sensitivity to get down to planets twice the size of Earth." So he wasn't quite looking for Earths; he was looking for super-Earths—a category of planet that nobody had thought much about. In our solar system, the next biggest planet up from Earth is Neptune, about four times as big across, and about seventeen times as massive, as Earth. A planet only twice as big as Earth might be similar in composition to Neptune, with a core of rock and ice surrounded by gases, or it might be mostly rock. If it's the former, it would be hard to imagine life on such a world; if it's the latter, maybe not so hard.

Or it might be something else entirely. From the moment the very first exoplanet was discovered, in 1992, astronomers have been ambushed over and over by the universe. The planets they find are not the planets anyone anticipated. The careful reader will assume "1992" is a typo that managed to sneak into print, since Michel Mayor's first universally acknowledged exoplanet, 51 Pegasi b, was discovered in 1995.

What happened in 1992, however, is hard to classify. Everyone agrees that Penn State astronomer Alex Wolszczan and Dale Frail, of the National Radio Astronomy Observatory, found two Earth-mass planets in that year. Yet the discovery was so odd, and fits so poorly into any sensible narrative about the search for a Mirror Earth, that you inevitably have to add a *but*. It's something like the old home run record in baseball. In 1927, Babe Ruth hit sixty homers, setting a record that went unbroken for decades. Then, in 1961, Roger Maris hit sixty-one. But Ruth hit his sixty in a season that lasted 154 games. By 1961, the season was 162 games long, and Maris didn't hit the last one until the final game of the season. So did he really break the record? This is the sort of thing baseball fans can argue about endlessly.

In the case of Wolszczan and Frail's discovery, the awkward part has to do with the star they were looking at. It's a pulsar, the collapsed, super-dense nugget left over after a star has died in the titanic explosion known as a supernova—an explosion so bright that for a few days it can outshine the rest of the stars in the galaxy combined. For a really big star, the leftover is sometimes a black hole. For something smaller, it's a chunk of matter just a few miles across, but more massive than the Sun. Just a teaspoonful of neutron star would weigh something like ten million tons. Some neutron stars rotate hundreds of times per second, and of these, some send out bursts of radio energy as they spin. When these incredibly rapid, precise blips were first picked up by radio astronomers in 1968, nobody had a clue what they were. They were briefly nicknamed LGMs, for Little Green Men, since at first nobody could think of a natural process that could cause such a precise, rapidly repeating signal.

What Wolszczan and Frail noticed was that the blips from a pulsar called PSR 1257+12 would vary slightly, coming closer and closer together, then farther apart, then closer. The best explanation was that something was pulling the pulsar toward, then away from the Earth. As the pulsar moves toward us, each blip of radio energy has just a little less distance to travel than the one before, so they come closer together in time. As it moves away, the blips have to travel a bit farther each time. The timing suggested two planets, each more or less the size of the Earth. It was crazy enough that Wolszczan and Frail held off on announcing the discovery, mindful of Richard Feynman's dictum that "you must not fool yourself—and you are the easiest person to fool." But while they were reanalyzing their data, a colleague named Andrew Lyne announced a planet around a *different* pulsar. Lyne's discovery was published in the prestigious journal *Nature* with great fanfare. Wolszczan and Frail figured they'd blown it. Maybe they'd been too cautious.

As it turned out, they hadn't been. A few months after his astonishing discovery had made headlines around the world, Lyne was preparing a triumphal talk about it for the upcoming meeting of the American Astronomical Society. Working late one night, he realized to his horror that he'd left out a crucial step in analyzing his observations. He knew, instantly and instinctively, what would happen when he redid the analysis. Sure enough, he told me shortly afterward, "the planet disappeared."

It would have been bad for Lyne's reputation if someone else had discovered the error before he did. It would have been even worse if he had insisted he was right after everyone else realized he was wrong. But Lyne found the mistake himself, and gave his talk as scheduled, but with a very different theme than he'd planned. The audience was utterly silent as he explained his mistake—and then, when Lyne finished, hundreds of astronomers gave him a standing ovation that lasted more than a minute. At the time, John Bahcall, Sara Seager's old adviser, was the society's president. After the talk, Bahcall came up to me and said, "I want you to know that Andrew Lyne's talk was the most honorable thing I've ever seen. A good scientist is ruthlessly honest with him- or herself, and that's what you've just witnessed."

Lyne's pulsar planet wasn't real, but Wolszczan and Frail's, it turned out, were. If the search for a Mirror Earth is ultimately a search for life, this discovery doesn't fit anywhere. Life couldn't possibly exist here. Planets couldn't survive a supernova explosion, so they must have formed afterward, out of some of the debris. But the neighborhood of a tiny, radiation-spitting neutron star would be about the most hostile environment possible. So the pulsar worlds are at once the first planets ever found and off at a tangent that makes them irrelevant to the search for a Mirror Earth. Exoplaneteers invariably mention the pulsar worlds when they talk about the history of exoplanet science, but don't talk about them much at all when discussing the current state of the research.

The pulsar planets were pretty much completely unexpected. So were the hot Jupiters Marcy and Mayor began to find in 1995. At that time, the theory of planetary formation was based on the assumption that our own solar system was probably typical. It wasn't obvious why this should be true, but it was even less obvious why it shouldn't be. The idea was that a cloud of interstellar gas and dust collapsed, less than five billion years ago, spinning faster and faster as it got smaller and flattening out into a disk. The dense core of the disk formed into the Sun, whose heat drove the lighter material, including hydrogen, helium, and water vapor, outward toward the edges and left only the heavier rocky material closer in. The rocky stuff formed into the Earth, Venus, Mars, and Mercury. Farther out, the rocky material congealed as well, but there was more of it. If you mentally divide the disk into bands, the bands farther out are much bigger around, so they have more total stuff—more rocky material, plus the extra gas pushed outward by the newborn Sun. The rocky matter would have congealed into big, rocky planets whose gravity then started vacuuming up the lighter gases. The result is worlds like Jupiter and Saturn, with solid cores shrouded in massive, thick atmospheres.

A Jupiter or a Saturn couldn't form close in, however, because there wouldn't be enough rock to form its massive core, and there wouldn't be enough gas to make the atmosphere. When 51 Peg b showed up, it was something like the situation when an elementary particle called the muon was discovered in 1936. No theorist had predicted such a particle, and the Columbia physicist Isidor Isaac Rabi responded by asking, rhetorically, "Who ordered *that*?" as though the muon were an exotic dish delivered unexpectedly from a Chinese restaurant. At least one theorist, however—Doug Lin, of the University of California, Santa Cruz—had predicted hot Jupiters. He'd argued, even before the discovery of 51 Peg, that in some cases a Jupiter-size planet could spiral inward, pushed toward its star by the gravitational effects of gas remaining in the disk. But most of his colleagues either didn't know about it or thought it was just a clever theoretical exercise. Suddenly, in the fall of 1995, it was a plausible explanation for a discovery that had come completely out of left field.

A pulsar planet could never support life. Neither could a hot Jupiter, not only because it's hot but also because any solid surface would be buried under the crushing weight of a thick atmosphere. For that reason, a cold Jupiter like ours wouldn't be likely to be habitable either, nor would a lukewarm Jupiter. If Lin's theory was right, however, the very existence of a hot Jupiter might rule out life anywhere in its solar system. If such a massive planet had spiraled inward, any smaller, Earth-like planet it met along the way would probably have been flung out of its stable orbit, and maybe even out of the system altogether. Hot Jupiters are the easiest planets to find, whether you're using radial velocities (they're big and close in, so they have the greatest possible leverage for yanking around their parent stars) or transits (a big planet blots out more light than a smaller one, and if it's in a tight orbit, it's more likely to pass directly in front of the star, purely by chance). The fact that so many early exoplanet discoveries were hot Jupiters could well be a biased sample. The first things you find are the easiest things *to* find. They're the low-hanging fruit that you can grab from the tree with the least possible effort.

If hot Jupiters are the rule, on the other hand, the odds of finding a Mirror Earth could be depressingly low. Within a few years after it launched, Kepler would presumably have an answer to this question. But the exoplaneteers weren't going to sit around waiting. "The probability of success is difficult to estimate," Philip Morrison and Giuseppe Cocconi had written in their *Nature* paper on SETI forty years earlier. "But if we never search, the chance of success is zero." They were talking about the search for extraterrestrial radio signals, but the same principle applied here. So while Borucki, Batalha, and the others on the Kepler team kept pushing ahead on building the spacecraft and the pipeline of software and human analysis that would turn raw observations into discoveries, everyone else in the business kept pushing on their own projects. No one could do better than Kepler—but everyone wanted to steal just a little bit of the mission's thunder.

Chapter 10

KEPLER SCOOPED

O F   C O U R S E   Y O U could never get to one meter per second." In 1999, when Debra Fischer codiscovered the second and third planet around Upsilon Andromedae, she was making what seemed to be a reasonable statement. Geoff Marcy had spent years struggling to convince other astronomers that his entire life's work wasn't a waste of time. Even after he and the Swiss team led by Michel Mayor began finding planets, their colleagues tried to argue that these weren't really planets, but something else. And even after it had become clear that they were planets and *not* something else, Marcy, Mayor, and the others who searched for wobbling stars—Bill Cochran and Artie Hatzes, of the University of Texas, for example, who had been looking for planets since the early 1990s; Gregory Henry, of Tennessee State University; George Gatewood, of the University of Pittsburgh; and many more—had to deal with the argument that they'd never be able to make measurements good enough to find planets anywhere near as small as Earth.

The problem was partly the precision of their instruments. "If you took a metal ruler a couple of inches long," Steve Vogt once told me, "and then stood it on end, the amount it would shrink due to gravity is the kind of effect you're trying to measure. If you picked it up, the expansion due to heat from your hand is a hundred times *more* than the effect you're looking for. And you've got to measure that." Marcy's iodine cell was one way to try to achieve maximum precision. Mayor's super-stable spectrograph was another. Both the California and the Swiss team had kept refining their instruments to push their precision even higher.

But they faced another problem as well. "I've been active in this field for twelve years now," Dave Charbonneau told me in 2010, "and I remember several times when people like Geoff would explain how they had worked so hard and improved their precision. And the naysayers would say—this is a good and important process in science, I want to make clear—the naysayers would say, 'I think you've hit the intrinsic limit of stars.' Stars have jitter, it is not a matter of having a better instrument, it is that the basic instability of stars isn't going to let you go below . . . and they would say a number. And that number would change over time and get lower and lower. Initially that number was five meters per second. 'You can't do better than five meters per second.' But the number kept going down."

It's true that all stars vibrate at some level, which makes the wobble from a planet much harder to tease out. "We know that some stars are intrinsically noisy," said Charbonneau, "and it would be very difficult to do these kinds of measurements. But there may be a healthy subset, maybe 15 percent, maybe 10 percent, that are extremely stable at a level of ten centimeters a second, which is where you have to get to detect an Earth. There's a lot of stars out there, so it could be that there's a sufficient number for finding Earth-like planets." This assumes that the spectrograph builders can create a device sensitive enough to detect them.

By the end of the 2000s, they hadn't built such a device, but they were getting closer. Michel Mayor's team in particular had built a spectrometer they called HARPS, for High Accuracy Radial Velocity Planet Searcher. In 2003, they installed it on a 3.6-meter-diameter telescope at the European Southern Observatory, at La Silla, in Chile. And even though it had been obvious to Fischer ten years earlier that you couldn't get there, HARPS was so stable, and so thoroughly well understood by the astronomers, that it had come all the way down to the one-meter-per-second barrier, and then broken it. Mayor was now down to half a meter per second.

This still wasn't sensitive enough to find a Mirror Earth, but it let the Europeans make a number of important discoveries through the decade, including several multiplanet systems. Perhaps the most intriguing of these, and ultimately the most controversial, made its debut in 2005. At first, the HARPS team was convinced only that they'd found one planet orbiting a star called Gliese 581 every 5.4 days. True to convention, the new planet was named Gliese 581 b. It was a hot Neptune, around seventeen times the mass of the Earth. But it wasn't all that hot, because Gliese 581, like many of the stars in the Gliese catalog, is an M-dwarf. Gliese stars are all relatively close to the Earth, and since M-dwarfs are the most common type of star in the Milky Way, it's not surprising that they're overrepresented in any catalog that simply samples everything within a given volume of space.

Since a bright, distant star and a dim, nearby star can look pretty much the same, Gliese created his catalog of nearby stars by looking for evidence of parallax—the apparent change of position of a nearby object when you look at it from a new point of view. You can do your own experiment to see how it works. Hold up one finger about six inches in front of your nose and close one eye. Then open that eye and close the other. As you alternate between one eye and the other, your finger appears to jump back and forth against the background. It does that because your eyes are a couple inches apart, so each has a different perspective.

Astronomers can see stars jump back and forth too, by looking at them at one time of year, then looking six months later, when the Earth has traveled through half of its orbit and is now on the other side of the Sun, about 186,000 million miles away from where it was. The nearest stars will appear to move against the background of more distant stars, even though they haven't really moved, because our perspective has changed (the more distant stars don't move, just as that tree in the distance didn't move when you blinked your eyes, because the change in perspective is small compared to how far away the distant objects are).

With a little bit of trigonometry, astronomers can use the change in viewing angle and the distance Earth has moved to calculate how far away the jumping, nearby stars are. Parallax is such an important tool for astronomers that they've invented a unit called the parallax-second, or parsec. It's how far away an object has to be in order to (appear to) move by just one second of arc, or 1/3600 degree, as the Earth makes half a revolution around the Sun. A parsec is about 3.26 light-years—and, in fact, astronomers almost always talk in terms of parsecs, or kiloparsecs, or megaparsecs, not light-years. The reason *light-year* is a more familiar term to most people is that parsecs take too long to explain, so astronomers convert for us.

Michel Mayor knew as well as Dave Charbonneau did that a planet would be easier to spot around an M-dwarf than around a bigger, Sun-like star, and that the habitable zone around the cooler M-dwarf would be closer in to the star. Mayor wasn't about to launch a radial-velocity version of Charbonneau's MEarth project (which in any case didn't even exist in 2005). He had enough to do already. But it wasn't crazy to include some M-dwarfs in the HARPS search. Gliese 581, an M-dwarf a little over twenty light-years away from Earth, or about six parsecs, turned out to be a gold mine. Once HARPS had found one planet orbiting the star, it made sense to keep watching. Since it was first, 581 b was by definition the easiest planet to spot in the system, but it wasn't necessarily the only one. The other planets, if they were there, might not be Earth-like, but a system with two or three or more worlds would at the very least help theorists understand how planetary systems form and evolve.

It took two more years of monitoring, but in 2007 those continued observations began to pay off. Mayor's team announced they'd found two more planets around Gliese 581. One was a world they'd already suspected was there. Labeled

Gliese 581 c, it was, they said, a minimum of 5.6 times as massive as Earth, which meant it was probably too small to have sucked in a smothering blanket of gases the way Neptune has in our own solar system. It might well be a rocky planet like Earth. It might even have oceans. The planet orbited once every thirteen days or so—quite possibly within the habitable zone of this cool, dim star. The planet's surface temperature, the astronomers calculated, was between 0° and 40° Celsius, or from just freezing to well below boiling. Water, if oceans existed, could be liquid. If Gliese 581 c wasn't quite a Mirror Earth, it was getting awfully close. "On the treasure map of the universe," team member Xavier Delfosse, of the University of Grenoble, said at the time, "one would be tempted to mark this planet with an X."

But in this case, it would have been wise to resist the temptation. How warm a planet is depends not just on how much energy it gets from its star, but also on what happens to the energy once it arrives. In our solar system, Venus is much hotter than Earth, and Mars is much colder. That's only partly because Venus is closer to the Sun, however, and Mars is farther away. It also has to do with their atmospheres. Venus is surrounded by a thick blanket of carbon dioxide, a heat-trapping greenhouse gas, which drives the surface temperature up to around 900° Fahrenheit, hot enough to melt lead. Mars has such a thin atmosphere that it retains very little heat; at best, the temperature rises into the twenties Fahrenheit. Billions of years ago, before Mars's relatively weak gravity let much of its original atmosphere escape, the surface was warmer, and liquid water flowed freely on the surface. We know this because orbiting spacecraft have seen unmistakable evidence of ancient river channels and lake beds, and because the Mars rovers *Spirit* and *Opportunity* have found minerals on the surface that almost certainly formed in the presence of water. When climate scientists ran their computer models on Gliese 581 c, they decided it wasn't likely to be habitable after all. Assuming it had an atmosphere, the planet was probably more of a Venus than an Earth, with a runaway greenhouse effect that would probably long since have sterilized it of any life that might have tried to take hold.

But there was a third world in the system as well, called 581 d, and over the next few years, the Swiss team would find still another, 581 e, and maybe 581 f (though they couldn't confirm this one). The Swiss had the best instruments—even their archrival Geoff Marcy admitted this—and they had been observing this star longer than anyone else; it's not surprising that they were the ones who kept finding new planets. But nobody gets to reserve a star to themselves, and while the Swiss had the best spectrograph, the instrument Steve Vogt had built for Geoff Marcy's team wasn't far behind. Once Mayor's rivals in the United States realized what a rich hunting ground Gliese 581 was, they began taking their own radial-velocity measurements as well. Maybe they could find a planet Mayor had missed.

Back in Europe, meanwhile, the European Space Agency had decided to beat Kepler into space with its own space-based transit mission. Just as Michel Mayor and his colleagues had no monopoly on Gliese 581, Bill Borucki had no monopoly on looking for transits from above Earth's atmosphere.

Mounting a competing mission as sophisticated and powerful as Kepler wouldn't be worth the trouble, since it couldn't be done significantly faster than Kepler, and it would be an expensive duplication of effort. So what the European Space Agency did was build a satellite less sophisticated and less powerful than Kepler, and get it into orbit as fast as possible. The satellite, named CoRoT (for COnvection ROtation et Transits planétaires) would be able to find planets only in relatively tight orbits around Sun-like stars, and it wouldn't be sensitive enough to find a planet as small as Earth. It could, however, find planets just a few times bigger.

In February 2009, just weeks before Kepler launched, it did. The parent star was known as TYC 4799-1733-1, TYC being the abbreviation for the Tycho star catalog. When the planet was spotted making a transit, CoRoT astronomers renamed the star CoRoT-7—they were free, after all, to create their own catalog, and the name would remind people that their satellite had made the discovery. They calculated that the planet, CoRoT-7b, was a bit less than twice the size of Earth. If that was correct, it would be, without question, the smallest exoplanet ever found. Like all transiting planets, it was also in the ideal edge-on orbit that would let radial-velocity instruments figure out its mass. Now that they knew where to point, Mayor's HARPS team swung their Chile-based telescope toward the star and began taking measurements.

Unfortunately, there was a complication. CoRoT-7 is similar to the Sun, but much younger—only about 1.5 billion years old, compared with our own star's 5 billion or so. Adolescent stars, like adolescent humans, don't always have the clearest skin. As Natalie Batalha had told me, they're prone to star-spots, and CoRoT-7 is no exception. This turns out not to be such a problem for measuring transits. Since the star rotates once every twenty-three days, the darkening caused by the starspots lasts much too long to be confused with a transiting planet. But spots can confound radial-velocity measurements. Depending on where the spot is at a given time, it can blot out part of the star's leading edge—the part that's rotating toward you—thus making it seem like the star as a whole isn't moving toward you as fast as it really is. Or it can blot out part of the trailing edge, with the opposite effect. CoRoT-7 is not the sort of quiet, middle-aged star that radial-velocity searchers liked to deal with.

Eventually, after a total of seventy hours' worth of observing time spread over the next several months, the HARPS astronomers managed to tease out a signal. "The longest set of HARPS measurements ever made," read a press release issued in September 2009, "has firmly established the nature of the smallest and fastest-orbiting exoplanet known, CoRoT-7b, revealing its mass as five times that of Earth's. Combined with CoRoT-7b's known radius, which is less than twice that of our terrestrial home, this tells us that the exoplanet's density is quite similar to the Earth's, suggesting a solid, rocky world." CoRoT-7b's composition might well be similar to Earth's, but its orbit is not. With a "year" lasting just a little more than twenty hours, it's more than twenty times closer to CoRoT-7 than Mercury is to the Sun, and has a surface temperature of between 3,300° and 4,700° Fahrenheit. If the planet was rocky, the surface might well be a sea of molten lava.

But that didn't take away from the potential importance of the discovery. Ever since Michel Mayor found 51 Peg b in 1995, exoplaneteers had been pushing toward smaller and smaller worlds, inching closer and closer to a true Mirror Earth. Nobody doubted that Earth-size, rocky worlds were out there, but it was possible that everyone was wrong. A universe where hot Jupiters could exist, in utter defiance of conventional astronomical wisdom, might also be a universe where rocky planets were vanishingly rare. Finding even one outside our own solar system would imply that they weren't rare at all. CoRoT-7b itself couldn't support life, but now it was clear that other small, rocky worlds must be relatively common, and some would surely turn out to be habitable. This was a major step forward.

It was, that is, *if* CoRoT-7b was truly made of rock. There's always some uncertainty in every astronomical measurement, because no measuring instrument—no telescope, no spectrograph, not even the best in the world—is perfect. A turbulent, spotty star just makes it worse. A longer series of observations can help, because turbulence is more or less random, while the back-and-forth radial-velocity tug a planet imposes on a star is like clockwork. Over time, that regular signal can build up to stand out from the visual noise caused by stellar turbulence. But the more noise there is, the more uncertain the signal will be, even with lots of observations. The CoRoT team's best calculation put the planet's density at about 5.6 grams per cubic centimeter, about the same as Earth's (water, by contrast, weighs one gram per cubic centimeter).

But other calculations—by other astronomers, and even by

some of the CoRoT scientists themselves—came up with densities ranging from much lower than Earth's to much higher. "I think I can say this without prejudice," Geoff Marcy told me. "Everybody agrees now that its mass is very poorly known. In fact, it's so poorly known that's its existence has been called into question. If I had to bet, I'd bet that something is orbiting with that period, but I don't know what that something is. And what its density might be is very hard to know."

The clearest illustration I've seen of this point appears on a slide used by several different astronomers at talks over the years. It's a plot of planet size on one axis and mass on the other. The chart shows the points where Earth, Neptune, and other solar system planets fall. But it also shows a line, passing through each of those points, where planets of similar composition, but greater or smaller mass, would lie. If a rocky planet like Earth were four times as massive as our home world, say, the plot shows how physically large it would be. Not a whole lot larger, it turns out, because a planet's mass increases much faster than its radius. One reason is that a more massive planet would squeeze itself tighter under gravity than Earth does. A rocky super-Earth would be denser than the actual Earth. But the more important reason is simple geometry: If you double the radius of a sphere, you increase its volume eightfold. Triple the radius, and the volume goes up by a factor of twenty-seven.

You can also take an actual exoplanet whose size and density you know and see where it fits on the plot: If it falls on the "Earth" line, it's almost certainly rocky. If not—depending on how far off it is, it might not be. If the planet's mass or size isn't known precisely, however, its density won't show up as a dot, but rather as a blob that covers the range of sizes and densities it might have. When you plot all of the possible combinations of size and mass for CoRoT-7b, based on all of the different estimates by different teams of astronomers, you get a blob that just barely edges over the Earth line. So it's possible that CoRoT-7b is a rocky planet. But there's a very real possibility that it isn't.

Whatever it's made of, CoRoT-7b is almost certainly a super-Earth—bigger than our home planet, but not as big as Neptune. It's not the first one ever found; a handful had already been discovered via microlensing, radial-velocity searches, and pulsar timing. CoRoT-7b is the first one, however, that could be described as even plausibly Earth-like in composition. It was the same sort of planet Dave Charbonneau had begun to search for with his ground-based MEarth project, and the same sort that Kepler should find easily. "There aren't any super-Earths in the solar system," Dave Latham told me during a visit to Cambridge. "There's a big gap. But," he said, smiling at what the CoRoT team had found, "old Mother Nature knows how to make them. It's a really nice discovery."

Despite the ambiguity about its mass, CoRoT-7b had taken the first step into the super-Earth era in exoplanetology. Later that same year—after Kepler's launch, but before the probe had announced any discoveries—the MEarth Project took the second. In May 2009 Dave Charbonneau was in a fancy hotel in Washington, D.C., getting dressed. He'd gotten a prize from the National Science Foundation—the Alan T. Waterman Award, given to an outstanding scientist under the age of thirty-five—and he was getting ready for the award dinner over at the State Department. "I don't swim in these D.C. waters," he told me later. "I'm a Boston academic and these people talk differently and dress differently. They put us up at the Ritz-Carlton, and I'm getting into a tuxedo, and I'm kind of nervous about what I'm going to say." He decided to check his e-mail, partly as a diversion and partly because he was missing the weekly MEarth meeting, when team members caught each other up on what they were doing.

MEarth had been under way for several months now. It was essentially a scaled-up version of the telescope he and Tim Brown had used to find HD 209458 b from a parking lot in Colorado a decade before. Instead of one telescope, MEarth had eight automated telescopes, each with a sixteen-inch-wide mirror, located on Mount Hopkins, near Tucson, Arizona. "They're essentially high-end amateur telescopes," he told me. Each one cost about $80,000, but, he said, "there are lots of amateurs who have invested eighty thousand dollars in their habit, often much to the chagrin of their partners." The MEarth group was made up of Charbonneau himself, two grad students, two postdocs, and an extremely earnest and serious undergraduate named Lauren Weiss, who has since gone on to grad school at Berkeley and to working with Geoff Marcy. "It's small enough," Zach Berta (he's one of the grad students) once said, "that we can all fit around a single table at a cheap Chinese restaurant."

Most of the team's weekly meetings actually took place over pizza at the Center for Astrophysics in Cambridge (Dave Charbonneau would send out e-mail invitations that began "Dear

Exoplaneteers," which is almost certainly where the term first arose). The meetings were largely devoted to going through possible candidates and weeding out false-positive detections. MEarth's biggest problem, just like Kepler's, was eclipsing binary stars that looked on first glance like transits. The MEarth pipeline was something like the Kepler pipeline, but, because MEarth was looking at so many fewer stars, and was looking at each one for a much shorter time, it was a lot simpler. If there were any actual transits in the data, these would be discussed here—or, if a team member was away from Cambridge, via e-mail. So before he tied his bow tie, Charbonneau logged in to see what might be new. "I'd been copied on an e-mail," he said, "from Jonathan Irwin [one of the postdocs] to Zach Berta, saying, 'Nice shooting, Zach.' There was a plot of a light curve attached, and as soon as you looked at it, you could see that there was this beautiful event." MEarth had found its first planet. "It was really fun to go to the dinner after that," said Charbonneau, "because I knew that this risky project we'd taken on had delivered after just six months of operation. And Dave Latham was there, so I got to confide in him that we'd found this really exciting planet."

The discovery, which the always-quotable Geoff Marcy would publicly declare to be a "top-of-the-top discovery in the quest for Earth-size planets," was a planet less than three times the diameter of Earth, according to the team's calculations. Its mass, determined by the European HARPS spectrometer, was about 6.5 times that of Earth. Nominally, it was a little bigger than CoRoT-7b, but observations of this new planet, called GJ 1214 b (the GJ prefix means the star appears in a version of the

Gliese catalog later expanded on by the German astronomer Hartmut Jahrheiss), weren't afflicted by the kind of uncertainties that clouded CoRoT-7b. When you do the calculation, the planet's density comes out at about 2 grams per cubic centimeter. If it were a blob of pure water, the density would come out at exactly 1. If it were perfectly Earth-like, it would be around 5.5. Neptune comes in at a little over 1.6. "So GJ 1214 b is in this funny place," Berta told me later, "in this weird intermediate place between Earth and Neptune, where it's less dense than Earth, but more dense than Neptune. It's difficult to understand."

Geoff Marcy described one possibility in an article he wrote for the journal *Nature*, in which GJ 1214 b was formally announced the following December: "It is likely," he wrote, "that this new world has nearly 50% of its mass in water surrounding an Fe/Ni [iron-nickel] core and a silicate [that is, rocky] mantle." That being the case, he wrote, "it probably has an extraordinarily deep ocean." Because it's so close to its star, with a year only 1.6 days long, GJ 1214 b is hot. But since the star itself is a dim red dwarf, it isn't *that* hot. The surface temperature would be about 320° Fahrenheit, well above the boiling point of water. So, wrote Marcy, "a sauna-like steam atmosphere is possible." The ocean would be more than a thousand miles deep, however, so the pressure at depth would be enormous, keeping most of GJ 1214 b's water in liquid form. It's pretty much impossible to imagine life existing on the molten-lava surface CoRoT-7b. It's not quite impossible with GJ 1214 b.

"What you want for life," Charbonneau told me at the time, "is a nice toasty ocean with a little bit of atmosphere. That's not going to be happening here. I think it would be foolish to say categorically that GJ 1214 b doesn't have life. But we have no basis for thinking it could." Jack Lissauer, a theorist at Ames and a Kepler project scientist, agreed. "We sometimes think of Earth as a water world, but we're not. If you have oceans hundreds of kilometers deep then you don't have little warm ponds." This was a reference to Charles Darwin's speculation that this is where life began on Earth. "I'd be very concerned," continued Lissauer, "if we found that most of the planets the size of Earth are water worlds. Of course, if you found signatures of life on them—well, you change your view. But you've got to do what you can with what you know."

Chapter 11

"A 100 PERCENT CHANCE OF LIFE"

BILL BORUCKI HAD said that the original Kepler team toasted with champagne when NASA finally approved the project in 2000 after so many rejections. By the time the satellite launched in 2009 it had taken nearly as long to go from approval to launch as it had from the first proposal to the last, successful one. That's how painstaking and careful the scientists had been in designing the mission, and how careful the engineers at Ball Aerospace, who put together the actual satellite, had been at building Kepler. The telescope was originally supposed to go up in October 2008, but the project took longer to finish than anyone expected for all sorts of reasons, so the launch was delayed until the following March—even more reason to celebrate when it finally happened.

Even up until the last minute, of course, Borucki was biting his nails, at least metaphorically. "Everyone," said Borucki, "was thinking of CONTOUR." This isn't exactly a household name, but in the space community, CONTOUR, the COmet Nucleus TOUR Mission, which was supposed to visit at least two and maybe three comets, was a well-known cautionary tale: It failed shortly after launch back in 2002 (by far the best Google result you get when you search for *Contour* and *spacecraft*: "UFO Destroyed Contour Spacecraft"). "In the back of my mind," Borucki said, "I was imagining this thing going up . . . and then going *plunk!* into the ocean."

But the spacecraft didn't fail. "It was a night launch," Borucki said. "It was beautiful. It was just so wonderful after all those years. To know we're now going to get data instead of just trouble. You can't imagine the elation we felt." So was there a second champagne toast? "Not exactly," he said. Kepler went into space atop a Delta II rocket from Launch Complex 17B at the Cape Canaveral Air Force Station. "It was a government thing," he explained, "so there couldn't be any alcohol at the launch or on the bus. We went back to a hotel and that was where we had our party, and I think there's a picture somewhere of me and Dave Koch drinking champagne. That was exquisite. I think we even had Dave dancing up on a table or something."

The launch was clearly a crucial milestone. Even so, Kepler wouldn't be announcing its first five planets until the following January. First, the spacecraft had to be checked out, during a two-month commissioning phase where engineers tested all of the electronics and other hardware to make sure it was working in space exactly as they'd planned. In practice, a spacecraft never does work exactly as planned, but knowing the difference between expectations and reality—that this light-detecting CCD, for example, is a few percent less efficient than the one next to it—lets the astronomers correct for any hardware or software errors. Real science observations began in May. Even so, however, it would take time for the software pipeline to identify Kepler Objects of Interest, and even longer to weed through them for planet candidates, and even *more* time to go out and confirm at least some of those candidates with radial-velocity measurements.

Those first five, moreover, which would finally be announced at the American Astronomical Society's 2010 Winter Meeting in Washington, would all be as big as or bigger than Neptune. It would take even longer for Kepler to get down to the level of a Mirror Earth, or even a super-Earth. Extracting a good signal from a transiting planet in the face of a star's roiling surface and flares and sunspots and weeding out electronic noise from the detectors required many, many observations. So even after Dave Charbonneau formally announced the discovery of GJ 1214 b, eight months after the launch, there was time for other groups to try to find the very first potentially habitable planet. GJ 1214 b wasn't the first; it was too wet. CoRoT-7b wasn't the first; it was too hot.

The Gliese 581 system, however, was proving to be just as rich a planet-hunting ground as Michel Mayor's team had first begun to realize back in 2005. The Swiss team had found four planets orbiting the star—planets b, c, d, and e. Planet c had looked like a good place for life, but it turned out to be too hot, so it was ruled out. But early in 2011, d, which had seemed clearly too cold, was belatedly ruled in. The problem with 581 c had been that any atmosphere it might have would likely trap too much of the star's heat, just as Venus traps the Sun's heat. A runaway greenhouse effect would, according to climate modelers who simulated the planet with a plausible, carbon-dioxide-rich atmosphere, make it too hot to be livable.

But then the modelers turned their attention to 581 d, a planet whose minimum mass was about seven times that of Earth. So 581 d was plausibly a super-Earth, even better than the real Earth at holding on to an atmosphere. If that atmosphere were rich in carbon dioxide—just as plausible for this planet as for its brother 581 c—the greenhouse effect would work in its favor. The modelers had more good news. Originally, the astronomers had realized that Gliese 581 d was almost certainly tidally locked to its star, showing just one face to the star at all times (just as the Moon shows only one face to the Earth). That implied that the perpetually dark side would be so cold that the atmosphere would freeze and collapse. The climate modelers showed, however, that this wouldn't happen: The atmosphere would circulate enough from the warm side to the cold side that the temperature differences should even out.

It was plausible, then, that Gliese 581 d was habitable. Without a transit to confirm its size and density, and maybe even to detect its atmosphere, there was no way of knowing. Maybe it was significantly more massive than its tug on the star would suggest. Even if it wasn't more than seven times as massive as Earth, Charbonneau's discovery of GJ 1214 b more than a year earlier had shown that a planet you might expect to be rocky and Earth-like based on size alone could instead be a water world, or even something more like Neptune than Earth—a rocky or icy core surrounded by a crushing, poisonous, hydrogen-rich atmosphere.

Mayor's team was not the only one looking at this star, however, and by the time Gliese 581 d had sneaked, barely, and only possibly, into the habitable zone, yet *another* planet had shown up in the Gliese 581 system. In September 2010, reporters got an e-mail alerting them to a press conference at the National Science Foundation where a major announcement about exoplanets would be made. Geoff Marcy's long-time collaborators Paul Butler and Steve Vogt were to be the speakers. When reporters saw the details, it was clear that the word *major* was justified. By combining their own observations with data from the Swiss team, Butler and Vogt had teased out the signal of two more planets, Gliese 581 f and Gliese 581 g. Planet f had about the same minimum mass as d, or about seven Earth masses, but was unambiguously outside the star's habitable zone. Planet g, however, had a minimum mass less than four times that of the Earth—and it was smack in the middle of the zone. "We're pretty excited about it," Vogt told me when I reached him at Butler's house the night before the press conference. "I think this is what everyone's been after for the past fifteen years." I contacted James Kasting, of Penn State, for an outside opinion, and Kasting, who is considered the world expert on habitable zones, agreed. "I think," he said, "they've scooped the Kepler people."

At the press conference the next day, Butler and Vogt acknowledged that without a transit to confirm Gliese 581 g's density, there was no way to guarantee that the planet had a rocky surface for life to walk or crawl around on. It could in principle be like Dave Charbonneau's water world, GJ 1214 b. There was no ambiguity at all about its surface temperature, however; it would be somewhere between -22° and -10° Fahrenheit—without an atmosphere. With any sort of reasonable atmosphere, it would be warm enough, at least on some parts of the planet (like others in the system, it always turns one face to its star), for water to be liquid most of the time.

Reporters naturally wanted a better headline than that, however (recall how the AP's Seth Borenstein buttonholed Bill Borucki after another press conference to pry loose something headline-worthy), so they asked explicitly about the prospects for life on the planet. At first, the astronomers demurred. They weren't biologists, Vogt protested, and Butler said, "I like data," making it clear that he didn't feel comfortable speculating. He wanted only to talk about things he could actually measure. But then, pressed repeatedly by reporters, Vogt gave in, and said something he would quickly regret. "The chances for life [on Gliese 581 g]," he said, "are 100 percent."

The reporters loved this, naturally, and ran with it. But many of his colleagues were appalled, or even worse. Some talked privately about Vogt as if he were a clueless buffoon. I spoke with Vogt a couple of weeks after the event, and he was clearly embarrassed at having taken the bait. "I'm not good at these media things where you have to speak in sound bites," he said. "I have no knowledge of whether there's life on the planet, obviously. I was expressing a personal opinion, or really, just a personal speculation." It was informed speculation, he argued, based on the fact that life on Earth seems to thrive under a huge range of conditions. Bacteria, in particular, have been found in super-heated water inside geysers and near volcanic vents at the bottom of the ocean. They've been found on pools of super-salty water, and pools tainted with radioactive mine tailings, and in pockets of water within Arctic sea ice, and encased in rock a mile underground. "Life seems to find a way, given even the smallest of chances," he said. "I look at 581 g, if it's there, and it is a place where life has a lot of chances of having a foothold."

If it's there, he said. The problem, as he well knew by the time we spoke, was that many exoplaneteers had become convinced that it wasn't. "The first principle," Richard Feynman said, "is that you must not fool yourself—and you are the easiest person to fool." Peter van de Kamp had fooled himself in the 1960s into thinking there was a planet orbiting Barnard's Star and never admitted he was wrong. Andrew Lyne fooled himself into thinking he'd found planets orbiting a pulsar, but salvaged his reputation by discovering the mistake himself. Now, even without Vogt's hasty remark about life, a consensus was rapidly developing that Vogt and Butler had probably fooled themselves. "There are e-mails flying around. People are frankly aghast," one exoplaneteer told me. "When I read their paper," said another, "red flags popped up. I was literally blown away. I couldn't believe it."

The reason, this anonymous astronomer said, was that Debra Fischer had written a definitive paper a couple of years earlier outlining the standard technique for extracting the signals of multiple planets from a complex jumble of radial-velocity measurements. "The methodology was very clear," he said, "and the pitfalls you have to watch out for were carefully outlined." Butler and Vogt were two of Fischer's co-authors on that paper. They had even gone on to debunk another planet claim using that same standard technique. "Now here we are," said the exoplaneteer with the red flags, "and Butler and Vogt have ignored the Fischer paper. They're using techniques that have proven to be unreliable and actually dangerous to use."

All of these doubts were confined to the planet-hunting community until about two weeks after the press conference. Then Francesco Pepe, a member of Mayor's team, spoke up at a conference in Turin, Italy. The Swiss group had gone back and reanalyzed their own observations of Gliese 581 with the HARPS spectrograph, folded in the observations Vogt and Butler had made with the High Resolution Echelle Spectrometer Vogt had built for the Keck Observatory, and found . . . nothing. "This does not prove there is no planet," Pepe told me. "All it shows is that we cannot see the signal. It's really a matter of how you interpret the data, how much confidence you have. In this case, we think there has been a mathematical misinterpretation." An American exoplaneteer was much less charitable. "This is not just a wrong result, this is a result from somebody who has gone off some deep end, and, what do you do? It's not to the advantage of the scientists who know the story to tell it. There is no value in it. It looks bad. Everybody loses, everybody from the scientists to the funding agencies to the individual scientists themselves loses if it looks like there was shoddy work being done."

There was a deeper issue, however, than whether Vogt and Butler had used the right statistical techniques. In 2007, after more than fifteen years of collaboration and an extraordinary track record of discovering hundreds of planets together,

Marcy and Butler had gone their separate ways, with Steve Vogt joining Butler in an independent venture they called the Lick-Carnegie Exoplanet Survey. This sort of realignment of research teams is often perfectly amicable. In this case, however, according to people familiar with the situation, it wasn't. Nobody involved in what astronomers have described as a "horrible breakup, like the worst divorce" was willing to talk about it on the record, but it was evident from a number of conversations that the major problem was clashing egos. Butler and Vogt were always seen as junior members of the team, with Marcy getting most of the attention, especially from the press.

There was, in short, no love lost between Marcy on the one hand and Butler and Vogt on the other. For that reason, Marcy and his current collaborators had to be careful about piling on to their former partners. "The intensity of what's going on is very harsh and negative," Debra Fischer told me. "These guys have a track record better than what the blogs are suggesting, but the penalty for being wrong in this business is extraordinary." Still, she said, "if Geoff or I say anything about Gliese 581 g, people might just put it down to the ugly breakup and bad karma between the two groups." When word got around that John Johnson, a new member of Marcy's collaboration, was writing a paper debunking Gliese 581 g, Fischer told me, "I had a Skype session with him telling him he shouldn't do it." Any takedown of the Vogt-Butler discovery would better come from outside the radial-velocity community entirely. It should be done instead, she said, by astronomers who specialized in statistics, like Eric Ford, of the University of Florida, or Phil Gregory, at the University of British Columbia. Mindful of the overlapping rivalries between the Swiss, Marcy, and Vogt-Butler teams, I realized I needed to go outside that community myself for an independent assessment. When I asked Steve Vogt whom he trusted for such an assessment, he, too, brought up Ford.

Eric Ford was a graduate student at Princeton in the early 2000s, a young, cheerful guy you probably wouldn't have pegged as someone who would become prominent. Ed Turner, a senior astrophysicist at Princeton, knew better. "It was quite obvious that he was exceptional even then," Turner said. Ford's primary research at Princeton was on something called the Space Interferometry Mission, or SIM. It was a complex space telescope NASA was cooking up to try to measure the side-to-side wobbles a planet would induce in its star if a solar system presented itself to us face-on. It was these side-to-side, or astrometric, wobbles that Peter van de Kamp thought he saw in Barnard's Star, and which NASA had talked about searching for before Bill Borucki and Geoff Marcy began their own planet-detection projects. "Probably two thirds of my thesis was on thinking about how you'd design a planet search for SIM," Ford recalled. Ford's thesis adviser was Scott Tremaine, the head of Princeton's astrophysics department at the time, and later John Bahcall's successor at the Institute for Advanced Study across town. Unfortunately for Ford, Tremaine went on sabbatical for a year in the middle of the research. "It was a sort of scary experience," Ford said. "Scott would sort of give you some ideas. You'd start to work on them, and suddenly he's off . . . I think he was hiking the

Adirondacks or something . . . some mountain range, I forget. And you sort of get to a point where like, okay, I'm really not making that much more progress. Instead of beating my head against a wall, maybe I'll start something new. So, I kind of made up something."

Six months later, Tremaine came back. "I go, 'Well, hi, Scott, I've been working on this other project. I hope you like it.' It made me a little bit nervous," said Ford, "but fortunately, he did." This other project was an attempt to come up with new statistical methods to analyze radial-velocity signals from multiple-exoplanet solar systems—exactly the problem Debra Fischer had solved for the Upsilon Andromedae system, and exactly the problem, except far more complicated, with Gliese 581. "So, it actually worked out amazingly well. Sometimes, either a little bit of luck or sort of looking for something where you say, 'What can I do that's different, where is the unique opportunity for me in this field?' Amazingly, it sometimes pans out."

By the time Vogt and Butler made their announcement, Ford was still young, energetic, and cheerful, but now he was also verging on eminence. When I asked him about the controversy over Gliese 581 g, he asked me in turn whether this was for a news story or for something more substantial. If it was for a story, he didn't want to comment. "One single quote is going to be misleading no matter what I say," he told me. I reassured him that it wasn't for a news story, and he went ahead.

"Okay, we're human," he began, carefully and gently. "We make mistakes. Sometimes we're under pressure to get results so we have a better chance of getting grants." In principle, he thought the paper's referees—the scientists a journal consults to see if a result is worth publishing—should have been skeptical about the Vogt-Butler paper all along. "But you can get a busy referee," said Ford, "or someone who doesn't have the relevant knowledge or experience to evaluate it properly." If the claim is boring, nobody might ever notice. "But if you make an exciting claim, you can be pretty sure someone will check on it."

Ford paused. Then he said, choosing his words very carefully, "I think the current data do not support the claim for this planet to a level of significance that meets the standard. I suspect that the analysis was not done as carefully as it could have been. I was skeptical as soon as I read it. It's a little bit unfortunate, because I'm obviously aware of the history [of the Marcy-Butler-Vogt breakup]. It may be that they're good scientists and do good work, but, in this case there was one part of the analysis where maybe they weren't as strong as when they were part of a larger team." But whatever the merits of Vogt and Butler's claims, he said, "I'm confident with time, science will figure out what's going on. It's a fascinating system whether or not this planet exists. The process will play itself out and come to a consensus."

Six months later, Phil Gregory, the second statistical expert Fischer had invoked, took a new set of observations from the European HARPS spectrograph, added them to all known observations of Gliese 581 from both the Europeans and the Americans, and analyzed them all. "I don't find anything," he said in an interview with *Wired* magazine. "My analysis does not want to lock on to anything around 36 days [the period claimed by Vogt and Butler for Gliese 581 g]. I find there's just no feature there."

At last report, Vogt was still telling me, as he had from the beginning, that he'd give up on the planet if the evidence went against him. But nothing he'd heard yet, including Gregory's analysis, was enough to convince him. Most astronomers now believe that Gliese 581 g was nothing but a mistake.

Vogt and Butler didn't seem to care. They were standing by their planet.

Chapter 12

THE KEPLER ERA BEGINS

B Y JUNE 2010, even though Kepler had been in orbit
for over a year, the team had announced only a handful
of planets—five in its first presentation, nearly a year after
launch, and just two more in the months afterward. But that
didn't mean the satellite wasn't seeing anything. Plenty of the
stars in Kepler's field of view were flickering regularly, and
even when the false positives were thrown out, the pipeline
had spit out 706 planet candidates in the first three months of
observation alone. The Kepler team had released 306 of these
in June 2010—not fully confirmed planets, but candidates they
were pretty sure of. With NASA's permission, they held back
the remaining 400 for further analysis.

They needed permission because, as a publicly funded mission, Kepler results had to be made public as soon as possible.
The Kepler team got first crack, but eventually they had to be
made freely available to other astronomers who could comb
the data for discoveries of their own. "Obviously," said Geoff
Marcy of the sequestered four hundred objects, "we have good
candidates, and you can bet that the implications of some of

them are profound. But if you say something publicly, it had better be unassailable. This is an unprecedented technical challenge, and the people at NASA Ames are working their butts off eighteen hours a day to make sure we get it right."

It's not as though the 306 that did get released were boring. The paper Borucki and the Kepler team published was titled "Characteristics of Kepler Planetary Candidates Based on the First Data Set: The Majority Are Found to Be Neptune-Size and Smaller." For Geoff Marcy, this was pretty mind-blowing. "Up to 1995," he wrote in an e-mail, "I thought we might not live to detect even one planet, of any size. Incredibly, the past 15 years brought us over 400 planets, painstakingly found one by one, most being giants like Jupiter and Saturn, with a few Neptune-size ones. Now these early Kepler returns hint strongly of a vast reservoir of ever smaller planets, approaching a few times the size of Earth, if not smaller—over 300 of these planets, each with their coordinates in space. It's a planetary treasure map. You can point the Keck and Hubble at them, or even your home telescope. There is only one appropriate assessment of this technical accomplishment and the haul of prospective small planets: WOW!"

The deal Borucki and Batalha and the rest had worked out with NASA headquarters was that they'd release the four hundred sequestered planet candidates, and maybe more, at a press conference around February 1, 2011—no backing out this time. This didn't mean, however, that there wouldn't be any more news from Kepler until then. Within just two months of the June data release, a paper appeared in *Science* announcing two Kepler planets that had been confirmed in an entirely novel way. It was so novel, in fact, that nobody on the Kepler team had even considered it when putting the mission together.

The most reliable way to move an object from the "planet candidate" column into the "planet" column was to detect a radial-velocity wobble, with the same period as the transit Kepler had seen. The two techniques were entirely independent, and no conceivable false positive could show up in both kinds of measurement. The added benefit, as Brown, Charbonneau, Marcy, and Henry had demonstrated with the first transit detection of HD 209458 b, was that by combining the two you could calculate the star's density. But most of the stars in the Kepler field were too faint to make radial-velocity measurements possible in any sort of reasonable time, even with a huge telescope like the Keck, in Hawaii.

The detection of HD 209458 b, however, had gotten a handful of exoplaneteers thinking—not about a new planet detection technique at first, but instead about a mystery concerning this first transiting planet from back in 1999. When both Marcy's and Brown's teams calculated the density of HD 209458 b, they found it was lower than they would have expected. The planet, a gas giant like Jupiter, appeared to be puffed up in size. "If I recall correctly," said Matt Holman, an astrophysicist at the Harvard-Smithsonian Center for Astrophysics and the lead author of the *Science* paper, one day in his Cambridge office, "there was a paper at the time, exploring how to make the planet bigger than the model would predict." One idea was that the planet had an eccentric, elongated orbit. That would create tidal forces that would squeeze and stretch the planet, heating up the interior and making the whole thing expand. Since the tides would also tend to damp out eccentricity and circularize the orbit, something had to be interfering with that process—the gravity of a second, still-unobserved planet in the system, for example.

"There was a little section in the paper," he recalled, "saying if there was another planet that had these kinds of properties and if you observed transit times, they would vary by a few seconds." The gravitational tug from the invisible planet wouldn't just keep the visible planet's orbit elongated; it would also make the transits vary from perfect clockwork, beginning a few seconds earlier or later than you'd normally expect. "I thought, 'Hey, that's kind of neat,'" said Holman. "There's this little dynamical thing." It surprised him that the variation was as long as a few seconds. "My intuition said it should be much smaller," he said, "so I set up my own numerical integration to check—and they were right. So I got to thinking about, okay, how big would that transit-timing variation be for a range of different types of planets with different orbits and different masses."

This was just the kind of problem that intrigued him. Unlike Dave Charbonneau, his colleague down the hall, Holman didn't become an exoplaneteer to discover life in the universe. "That's not really what motivates me," he said. Instead, he explained, "I'm always motivated by precision. What's always excited me about planetary dynamics is that you can make very careful, detailed predictions and detailed measurements and you can write down the equations of motion and I like that." In the 1980s, the field of chaos theory was just emerging, and along with a handful of other planetary scientists at the time, Holman had tried to figure out whether the solar system is stable over the very long term. (The answer is no, but planets are stable enough for long enough that we don't have anything to worry about—unless an unstable body like a comet or asteroid changes direction and heads right for us. Which has happened before, and will almost certainly happen again.)

"What was really exciting about extrasolar planets," he continued, "was that suddenly there were a lot more systems to study." Thanks to Kepler, moreover, there were going to be lots of systems where the timing of transits could be clocked right down to the second. If there were any variations in that timing, it would allow scientists to deduce the mass of the planets—an entirely different way of doing so than through radial-velocity wobbles, and one that didn't depend on the star being especially bright.

Fortunately for Holman, the Kepler team had created a category of semi-insiders called Participating Scientists. They weren't formal members, but they were allowed to propose supplementary projects that would make use of the Kepler data. "It's really a grant program," said Holman, "where not only do you get the money to do the work, but you get the opportunity to be part of the Kepler team. I actually don't know whether that's done commonly or not." Sara Seager is a Kepler Participating Scientist (she's looking for reflected light from Kepler planets to try to characterize their atmospheres). So is Dave Charbonneau (using the infrared-sensitive Spitzer

Space Telescope to weed out false positives among Kepler Objects of Interest), Eric Ford (characterizing the eccentricities of Kepler planets), and several others.

So Holman wrote a proposal to look for transit-timing variations in the Kepler data, and it was accepted, and he and a group of colleagues began looking through the Kepler Objects of Interest as they came out of the pipeline. "I'm not looking at 150,000 or 160,000 things," he told me. "I'm looking at the ones where they have already said, 'We think that there are planets here and we've gone through the exercise of ruling out all the obvious other things they could be.'" The "they" in this case was mostly Jason Rowe, a postdoc on the Kepler team who was leading the vetting process. "He's the one that finds the objects of interest. He actually did the preliminary time measurements so we could see that there's more than one planet there and one looks like it's speeding up, one looks like it's slowing down. Once the transit-timing variation group saw that, we went, 'Okay, that's the one. We're going to really focus on that.'"

The system they focused on was originally known as KOI-377; by the time the paper was published in *Science*, it had been rechristened Kepler-9, signifying that it was no longer just a star with planet candidates, but with actual planets. (A word about the Kepler numbering system: The first five stars where Kepler discovered planets, announced months earlier at the AAS meeting in Washington, were called Kepler-4, -5, -6, -7, and -8. There was no Kepler-1, -2, or -3, however, since Kepler numbers go only to planets actually discovered by the Kepler Mission. The first three stars where Kepler spotted transiting

planets had already been discovered by ground-based telescopes. Borucki and his team had decided to ease into the
project with three detections that should be absurdly easy. If
Kepler couldn't see planets everyone already knew were there,
it would have been a very bad sign.)

In the case of Kepler-9, there were two transiting planets,
one with a nineteen-day orbit and the other at thirty-nine.
But the timing of those transits wasn't like clockwork: They
varied by four minutes and thirty-nine minutes, respectively.
Holman was the lead author on the *Science* paper that announced the new result, but he had more than three dozen co-
authors. They were all listed by name, but they simply have
been identified as the Exoplaneteer All-Stars, since they included Bill Borucki, Natalie Batalha, Dave Charbonneau, Eric
Ford, Geoff Marcy, Dave Latham, Debra Fischer, and others.

This discovery was a very big deal. Not only was it concep-
tually elegant, but it also allowed scientists to calculate, using
the laws of gravity, exactly how massive each of the planets in
the system was. Until now, the only way to get a clue about a
transiting planet's mass and density was to use the radial-
velocity-wobble technique to see how hard it tugged on its star.
For most of the stars in the Kepler catalog, this wasn't possible;
the stars were too faint. Transit-timing variations were a second
way into the problem, and the brightness of the star was irrelevant. You did need more than one transiting planet, obviously.

But the Kepler insiders knew something they hadn't yet revealed to the world: There were far more systems with multiple transiting planets than anyone had suspected. "It was a big
surprise," Dimitar Sasselov told me confidentially a month or

so after Holman's paper came out. Sasselov, who got his Ph.D. in Communist Bulgaria before immigrating first to Toronto and then on that well-worn path from Toronto to Cambridge, Massachusetts, was a full-fledged Kepler co-investigator, along with Geoff Marcy and Dave Latham and—until she was promoted to deputy principal investigator—Natalie Batalha. "We didn't expect it," he said. "We kind of had the feeling, 'Who ordered this?' You know? We expected maybe 5 systems with multiple transiting planets. We've found 170." He noticed a look of astonishment on my face. "Yeah, 170 multiple-transiting systems, many of them with double transits, but there are plenty of triple transits. One has six. One has five, and there are a few with four."

This was really more than a nonphysicist's mind can grasp all at once, but the second author on Matt Holman's *Science* paper, a Santa Cruz postdoc named Daniel Fabrycky, came up with a rather brilliant way to illustrate it. It happened by accident, really. When the six-planet system (it was dubbed Kepler-11) was finally announced early in 2011, Fabrycky, who would co-author that system's discovery paper as well, wanted to visualize what was just a bunch of numbers only a scientist could love.

"Someone had figured out," he explained, "that three of the planets would sometimes cross in front of the star at the same time, and I wanted to figure out where the other three were when they did." So he wrote a computer routine that would show where all the planets were at a given time. To advance the simulation one notch forward in time you'd press on the computer's F key. Knowing where all the planets were allowed NASA to create a vivid and accurate artist's rendering of what the system might look like if we could travel there and see it up close. The editors of Nature would end up putting the illustration based on Fabrycky's simulation on the magazine's cover.

But Fabrycky's son and daughter, ages six and four at the time, respectively, figured out something else. If you kept hitting the F key over and over, you'd turn the simulation into a crude movie of the planets going around and around and around. It was pretty entertaining. Fabrycky, who has since joined the faculty at the University of Chicago, also realized that his program would work equally well for all 170 or so of the multiple-planet systems as it did for Kepler-11. So he put together an animation showing all of these systems, as though seen from above, on a single screen, and set them in motion. Fabrycky titled his creation the Kepler Orrery, after the gorgeous mechanical solar systems built in the 1700s to illustrate the motions of the planets.

Then he added one perfect final touch. Since the animation was virtually buzzing with a swarm of planets, he created a sound track with Rimsky-Korsakov's "Flight of the Bumblebee," played on a xylophone. When he (or anyone else) flashes it on the screen during a talk, there's a moment of stunned silence while the audience takes it in. Then the room invariably erupts into delighted laughter.

Chapter 13

BEYOND KEPLER

WHEN GEOFF MARCY and Michel Mayor began finding planets in the mid-1990s, their discoveries triggered a wave of excitement in the astronomical community that was unlike anything the field had seen, probably ever. It was also immediately clear to everyone that the ultimate goal would be to detect life beyond Earth. But it was also clear that existing telescopes weren't nearly powerful enough to do that. As Bill Borucki had discovered back when he first started thinking about Kepler, NASA had been playing with ideas for finding planets, and even finding life, for decades. The agency had produced a number of reports and white papers on the topic, but hadn't done much more than that.

All that planning, however, wasn't entirely in vain. At the same meeting where Geoff Marcy announced his first two planets back in 1996, Daniel Goldin, then the NASA administrator, was able, thanks to those years of study, to step up to a microphone the day after Marcy's talk and lay out a fully formed, step-by-step strategy for identifying and then studying a Mirror Earth. The key technology would be something called interferometry, a technique for combining the light from two widely spaced telescopes to simulate a single, gigantic scope with extremely high resolution—the ability to take super-sharp images.

The first step, called the Space Interferometry Mission, or SIM, would use that sharp resolution to do astrometry—to measure the side-to-side wobbles a planet imposes on a star rather than the forward-and-back wobbles used in radial-velocity searches. Astrometry was so hard to do that both Bill Borucki and Geoff Marcy rejected it when conceiving their own projects. Nevertheless, said Goldin, SIM would do astrometry with such precision that it would be able to detect the wobbles induced by Mirror Earths as they orbited around Sun-like stars. (Goldin was also the NASA chief who declared when he took office that henceforth, the agency would do everything "better, faster, and cheaper." Scientists and engineers generally agreed that they could do any two of the three at one time, but not all of them.)

The next grand step would be the Next Generation Space Telescope, a successor to the Hubble. The NGST, which has since been renamed the James Webb Space Telescope, should be in orbit by 2007, said Goldin (the current best estimate is 2018). While the NGST wouldn't be designed just for planet-hunting, it might be able to take images of giant planets, as long as they were far enough away from their stars that they would be lost in the glare. Then, by 2020 or thereabouts, he said, the crown jewel of NASA's planet-searching program should be ready for launch. Called the Terrestrial Planet Finder, or TPF, it would be an interferometer like SIM, but with four huge space telescopes instead of two small ones. These four telescopes would have to be so widely separated that they couldn't sit on a single structure. They would have to fly in formation, out in the general neighborhood of Jupiter, maintaining their separation to within a fraction of an inch as they sailed through interplanetary space.

If it all worked out, the Terrestrial Planet Finder would be able to do something remarkable. By adjusting the spacing of the telescopes just slightly, interferometry would cause the light to cancel out in parts of the image. In principle, you could blank out the star, making it much easier to see an Earth-size planet, and even to probe its atmosphere for the chemical signature of life. This would be incredibly difficult, technically, but hadn't NASA landed men on the Moon, and set cameras down on Mars, and detected the faint afterglow of the Big Bang with the COBE satellite?

Despite the fanfare and the promise, however, this grand scheme didn't play out quite as Goldin had portrayed it. "In 1999," Marcy recalled, speaking more than a decade later, "the Space Interferometry Mission was approved, the budget was roughly $50, $60, $70 million per year. We met three, four times a year, the science team did—an enormous effort." Eric Ford's graduate thesis at Princeton was in support of the SIM program. NASA ultimately spent $600 million on the project without even starting to build the hardware, and then canceled it for budgetary reasons. The more ambitious Terrestrial Planet Finder hasn't been canceled, but it's been put on hold. If TPF launches by 2030, astronomers will be very surprised.

The problem with TPF isn't just that it's expensive and

technically difficult, but also that astronomers can't even agree on what the instrument should look like. The original concept involved those four space telescopes flying in formation, but in the early 2000s, designers came up with two simpler (though less powerful) alternatives. The first was to build a large, single space telescope and fit it with a coronagraph, a device that blots out the light of the central star to let a planet shine through. The second was like the first, but would involve the launch of two separate pieces of hardware: the telescope itself and a device called an occulter. The occulter would fly thousands of miles away from the telescope and position itself in just the right way to blot out a star. Each version had its proponents who argued that theirs was the right one and everyone else's was wrong. In the end, NASA threw up its hands, put TPF on the back burner, and cut the project's budget drastically.

Even in what amounted to suspended animation, however, the few scientists and engineers still working on TPF have continued to make progress. I was in Berkeley one evening for an observing run in the basement of Campbell Hall, the physics building, where Geoff Marcy was using the Keck II telescope to do radial-velocity follow-ups of Kepler candidates. When he and Paul Butler had first begun using the Keck, in 1996, they had to fly to Hawaii, make their way up to the telescope, in the cold, thin air at the summit of Mauna Kea, and try to stay awake as they fought jet lag and oxygen deprivation. Nowadays, with a super-broadband Internet connection, all of the adventure and romance is gone. I haven't yet heard anyone complain.

The observing run wouldn't start until close to midnight, so I was killing time in the hotel lobby when in walked David Spergel, the head of Princeton's astrophysics department and one of the original members of Princeton's TPF collaboration. He'd come out from Princeton for a dinner celebrating the career of his graduate school mentor, Leo Blitz. The dinner was over and he was headed for bed, but he had a few minutes to talk about his own work on TPF. "It's not completely dead," he said. "There's something like one hundred or two hundred million dollars in technology development available."

In fact, the Princeton TPF project, whose genesis had come during the conversations that engaged Sara Seager so deeply in the early 2000s, had turned into two parallel projects, one focused on the coronagraph idea and the other on the oc-culter, also known as a starshade. The first was located in the university's school of engineering, where the Princeton TPF group's principal investigator, Jeremy Kasdin, explained both projects during a visit by Michel Mayor in the fall of 2011. The coronagraph arm of the research, said Kasdin, was still using the "pupil" idea Spergel had come up with earlier in the decade, which didn't blot out starlight so much as shunt it off to the side of the image, letting a planet become visible.

The pupil had evolved from a simple cat's-eye shape to a mask with a set of elaborate, curved openings, which turn out to be even more efficient at shunting light away from parts of the image to let planets shine though. "How do you come up with these shapes?" asked Mayor. "Is it just intuition?" Not exactly, said Kasdin. "Usually, I'll get out of the shower with a great idea, and I'll call Bob." Bob is Robert Vanderbei, chair of

Princeton's Operations Research and Financial Engineering Department, and an expert in "optimization." That's a field of mathematics/computer science that lets you adjust anything from an investment portfolio to an airplane wing for maximum performance. Vanderbei is also an extraordinarily talented amateur astronomer and astrophotographer, so it's not surprising that he found his way into Kasdin's group. When Kasdin calls Bob, he's looking for help in refining a new idea for a pupil shape into something that reduces light in an optimal way.

The lab where Kasdin and his group tests out new pupil shapes is located inside the university's engineering building, while the lab where they test the second version of TPF, which will use a free-flying occulter, is about three miles away, on the university's satellite Forrestal Campus. Not far from the occulter lab is the Princeton Plasma Physics Laboratory, where scientists and engineers have been doing experiments in controlled nuclear fusion since the late 1950s. That bears mentioning because the man who founded the fusion-energy lab, Lyman Spitzer, was also the father of the Hubble Space Telescope, which he first thought of in the late 1940s. (Spitzer didn't get his name on the Hubble, but he's memorialized by the infrared Spitzer Space Telescope, which Dave Charbonneau now uses to look for planetary atmospheres.) Spitzer also worked on planetary-formation theory early in his career—and to top it off, he wrote an early proposal for free-flying occulters back in 1962.

The occulter Princeton is working on, in partnership with NASA's Jet Propulsion Lab (JPL), Kasdin explained, would be 40 meters, or about 130 feet, across; it would fly about forty thousand miles away from the telescope itself, and it would have to maintain its position to within about two feet. The occulter wouldn't be just a round disk, but would rather (because of diffraction, and thanks to optimization analysis) look something like a flower, with twenty stubby petals coming out of a wide central hub. The edges of the petals have to be as sharp as razor blades, said Kasdin; if they were any thicker, they might reflect too much stray light from the Sun into the telescope, tens of thousands of miles away. When I ran into David Spergel in Berkeley, he'd come not directly from Princeton, but from JPL, in Pasadena, where engineers had built a full-size mockup of one of the petals. He'd witnessed a demonstration of how the petal would unfurl in space. It would have to unfurl, since there's no rocket big enough to contain a hundred-foot-wide flower unless the blossom is folded up on itself.

Spergel agreed that TPF-I, the original four-telescope interferometer, was probably too complicated and difficult to get off the ground. It was basically four James Webb Space Telescopes, and even one is turning out to be very hard and very expensive to build. "The occulter is a lot easier," he said. "You're flying two objects, and you have to keep them aligned, and deploy the occulter properly, so it's still a lot of work. But the fact that I saw them deploy a mockup at JPL makes me optimistic."

I heard the same from Jim Kasting when we met a couple of months later at an American Astronomical Society meeting. Kasting has been an exoplaneteer for nearly as long as

Geoff Marcy, Bill Borucki, Michel Mayor, and Dave Latham, which is to say he's been at it since the late 1980s. He doesn't search for planets, and never has. Instead, he's perhaps the world's leading authority on habitable zones—the orbital bands around stars where water can be liquid and life therefore stands a chance of gaining a foothold. Like Marcy and the rest of the exoplanetology community, Kasting was under the impression that NASA was actually talking about funding not one, but two versions of TPF a few years ago. "There was a brief period of extreme optimism," he said, "where we thought we were going to get two big flagship-type missions. Somebody was crazy, basically." (*Flagship* is the term reserved for complex, expensive missions that cost a billion dollars or more.)

Then both of them went away. "TPF has been off the drawing boards, basically nowhere, for the last five years, and then it resurfaced as a technology-development project," he told me one evening over dinner at an inexpensive, minimally decorated, and excellent Vietnamese restaurant a couple of blocks from the Seattle waterfront. "They actually put technology development for TPF at the top of their medium-priority list. We were all disappointed that there wasn't a dedicated exoplanet mission. That hurt a lot of people who had been working on it for fifteen years or more. But we all see the bright side of it, where if we play our cards right, some form of TPF will be the next big flagship mission."

The flagship space mission the Decadal Survey, a once-every-ten-year report from the astronomical community to NASA, did recommend was something called the Wide-Field Infrared Survey Telescope, or WFIRST. It would mostly do cosmology, including research on the mysterious dark energy that appears to be making the universe expand faster and faster as time goes on. "It has a small exoplanet component to do gravitational microlensing, but aside from that, most of us are not that thrilled about it," Kasting said. "But," he continued, "they're going to be lucky to ever fly that thing."

The reason is the James Webb Space Telescope, aka the Next Generation Space Telescope, the successor to the Hubble that NASA has been thinking about since the mid-1990s. Originally, the Webb was supposed to have a light-gathering mirror eight meters, or more than twenty-six feet, across. That proved too expensive and too hard to launch, so the mirror was eventually downsized to six and a half meters. That's still pretty big: the Hubble's mirror, by comparison, is less than two meters across.

The original launch date was supposed to be 2007, but that was never really firm. The actual launch, once NASA started funding the project seriously, was set for 2015. In the fall of 2010, however, an independent review panel requested by Congress reported that the Webb project was badly behind schedule and over budget. Under the best of assumptions, there would be no launch until 2018 at the earliest, and the money it would take to do that would come at the expense of other projects. "This is NASA's Hurricane Katrina," said Alan Boss, a planet-formation theorist at the Carnegie Institution of Washington, to the *New York Times*. It will, he said, "leave nothing but devastation in the astrophysics division budget."

This could actually work to TPF's advantage, however, according to Kasting, "because the Europeans will probably fly their Euclid satellite, which does a lot of the same science as WFIRST. Many of us think it's a waste to do both WFIRST and Euclid." If Euclid flies late in the current decade, and if WFIRST gets pushed back by Webb's budget problems, maybe WFIRST will get canceled. "So that could help us. If WFIRST went away then maybe the next flagship could be TPF. Thinking optimistically," he added.

But if Alan Boss is right, there's also a realistic possibility that the Webb telescope could take money not just from future missions, but that it could also hurt Kepler. Originally, NASA agreed to a four-year Kepler Mission, with the possibility, but no guarantee, that the agency would spring for another four years, in what would be called an extended mission. Like the Webb, Kepler went over budget, albeit far less drastically, so the original mission was cut to three and a half years. "We're up for senior review on the extended mission in February 2012," Natalie Batalha said, "and I'm really worried." She was even more upset when, in the spring of 2011, a House committee voted to cancel the Webb entirely. "JWST is just a double-whammy. The whole community has sacrificed to fund it. Everyone was unhappy at how much it was costing, but we knew how valuable it could be. And now you have Congress talking about canceling it."

In the end, the Webb wasn't canceled. Barbara Mikulski, the Maryland senator who runs the Senate Appropriations Committee, is a big supporter of the telescope, at least partly because its headquarters, like that of the Hubble, is in Baltimore. The Goddard Space Flight Center, from which the Webb will be controlled, is in Greenbelt, Maryland. She put the Webb back in the Senate's version of the NASA budget, and that's the version that survived. It's a good thing for astronomy, if not for planet-hunting in particular: The Webb will be one of the most powerful astronomical instruments ever built, able to peer all the way back into the Dark Ages shortly after the Big Bang, when the stars first began to form. But the Webb will be useful for exoplanetology too: Its infrared-sensitive detectors will be able to analyze the reflected light from hot Jupiters and hot Neptunes to see what their atmospheres are made of—much like what Dave Charbonneau and others have been doing with the Spitzer and Hubble space telescopes, only more effectively. You could even fly an occulter along with the Webb, making it an ad hoc version of TPF.

Like the Spitzer and the Hubble, however, the Webb is a general-purpose telescope, where Kepler has an extremely narrow mission. Any planet-hunting duties would have to compete for Webb observing time with all of the other science the telescope is capable of doing. "I'm worried," said Batalha, "that Kepler will make this amazing catalog of planets and there won't be anything to follow it up with—that we'll end up waiting decades and decades to explore these worlds we've found."

That might well be true of the planets Kepler identifies in any case. As everyone knew from the beginning, most of the Kepler stars were too faint to follow up transit detections with radial-velocity measurements. For Earth-size planets in the habitable zone, there was pretty much no way it could happen. And if you did manage to confirm such a planet through transit-timing variations, say, you'd still need a TPF or a Webb-plus-starshade to have a hope of studying it.

For many young exoplaneteers in grad school or doing postdocs or holding junior faculty appointments, the litany of canceled and postponed space missions is clearly discouraging. Still, a few missions, less expensive and less ambitious but still potentially exciting, continue to move forward. One of them is being cooked up in the Green Building at MIT—the same place Sara Seager works. It's a high-rise, the tallest building on the campus, topped with two white, spherical radar domes that have been there, looking down on the Charles River Basin, at least since I was in college back in the early 1970s. (As a sophomore, I had a job driving the motorboat for the freshman crew coach. On cold November afternoons, the sight of those domes, still lit by a Sun that had fallen below the horizon from where I was sitting, was a reminder that it would be at least an hour before I could get back upriver to the warm boathouse and stop shivering.)

I heard about the mission from Josh Winn, an affable young assistant professor who began his astronomical career, much like Dave Charbonneau and Sara Seager, in cosmology. As an undergraduate at Princeton, he worked on gravitational lensing, trying to measure the size of the universe by looking at the flickering of quasars. The idea is that when the gravity of a nearby galaxy distorts the light of a distant quasar, it can create a multiple image—what looks like two or three or even four quasars where there's actually just one. The light paths the images follow to our eyes vary slightly in length, so when the actual quasar flickers, the flickering shows first in one image, then in another. This time delay (plus a bit of calculating) tells you how far away the quasar really is. "When I talk to people about lensing," Winn said, "they listen politely, but mostly their eyes glaze over." Now that he's part of the search for life on other planets, he said, "they get it right away. I don't have to explain why it's important."

This isn't the main reason Winn became an exoplaneteer, but as he went through grad school at MIT he felt the urge to do something a little more practical. He tried medical physics, but it didn't click with his personality, so he returned to lensing, and also did some work in condensed matter physics. After grad school, he did a stint as a science journalist, writing for the *Economist* for a year, and he considered abandoning science in favor of writing. In the end, he took a postdoc at Harvard, where he continued to work on cosmology. But just in the years since Seager and Charbonneau had departed, Harvard had become a hotbed of exoplanetology.

"There was all this excitement in the air about exoplanets," said Winn. "The prospect was just emerging that we could study their atmospheres, and there were all of these other intriguing physics problems posed by multiple planet systems and close-in planets." Back at Princeton, meanwhile, his old mentor, Ed Turner, along with many other senior astronomers, had turned into exoplaneteers as well. "It seems to me," he said, "that the early days of lensing in the 1980s had that same feeling of excitement and newness as exoplanets. But exoplanets have more staying power because of the quest for life."

So now, among many other duties, Winn was serving as the project scientist for a space telescope project called TESS, the Transiting Exoplanet Survey Satellite. "It would be a successor to Kepler," he explained, "but looking across the whole sky rather than at one narrow area." The trade-off would be that TESS would gaze at each of the two million stars on its list for a much shorter time—months, not years. It couldn't find a Mirror Earth, with an orbital period as long a year. The biggest thing in TESS's favor is that if JWST or even TPF finally goes into operation, it will have a nice, juicy set of planets to look at: The stars in the TESS catalog are much brighter, on average, than the ones Kepler is looking at. That makes radial-velocity follow-up with a telescope like the Keck much easier. It also makes it easier to study the light passing through or bouncing off the planets' atmospheres, so exoplaneteers can study their compositions. And ultimately, if future telescopes can image the planets directly, the fact that TESS planets are closer to Earth will make those observations easier as well. It is, said Winn, "the natural next thing to do."

It's also far cheaper than a billion-dollar flagship mission. Kepler was a Discovery-class mission, limited to a budget of no more than $300 million. TESS was being proposed as a Small Explorer mission—a SMEX, in NASA's acronym-happy universe—which had to come in at under $200 million. The project's principal investigator, George Ricker, also at MIT, had first submitted a proposal to the agency in 2009, but it didn't make the cut. The team resubmitted at its next opportunity, however, in early 2011, along with twenty-two others;

the following September it was selected as one of five that would get $1 million for an eleven-month "concept study." The best two of these would go on to launch, as early as 2016. If TESS loses out on the final round, the team might want to call Bill Borucki in as a motivational speaker.

Chapter 14

HOW MANY EARTHS?

A S THE KEPLER team tried to remind reporters every time their satellite found a new planet, finding planets wasn't the goal—not individual planets, anyway. The goal was to determine how many stars, on average, have a Mirror Earth orbiting around them, a planet of about Earth's size, located in the star's habitable zone. If the percentage is high in the Kepler sample, that boosts the odds that there will be Mirror Earths close to us. If it's low, you need a flagship mission after all, which can look farther out into the Milky Way.

It all depends on a number exoplaneteers have begun, over the last few years, to call η_{Earth} (that's the Greek letter eta, so the term is pronounced either "eta Earth" or, more commonly, to make it clear that the word *Earth* is written as subscript, "eta-sub-Earth"). It's the fraction of Sun-like stars that have a Mirror Earth orbiting them—or that's one definition. "There are actually many different definitions," Andrew Howard, a postdoc working with Geoff Marcy, told me during my Berkeley visit. "That one is just the narrowest."

Sometimes, Howard explained, people use the term to mean

Earth-mass planets around Sun-like stars, without saying anything about the planets' sizes. Sometimes, as with Kepler, they
mean Earth-size, whatever the mass. Sometimes the definition is expanded to include M-dwarfs, not just Sun-like stars.
Ultimately, it doesn't matter all that much. The point is to
figure out how hard it will someday be to focus in on a Mirror
Earth and search for evidence of life. If eta-sub-Earth is around
10 percent, Jim Kasting says, and you're talking only about
Sun-like stars, that means you should expect to find just three
Mirror Earths within the nearest fifteen parsecs, or about fifty
light-years—about three hundred trillion miles in all directions. "I'm actually optimistic," he said, "that eta-sub-Earth is
going to end up higher than 10 percent. I think it's going to be
more like 20 percent to 40 percent, somewhere in there. But we
will have to wait for the Kepler folks to tell us."

He said this in the knowledge that Kepler was up and
working and beaming down information faster than the team
at Ames and their collaborators elsewhere could process it.
None of this had been certain back in 2007, when NASA officials came to Geoff Marcy urging him to write a proposal to
come up with a preliminary number for eta-sub-Earth from
the ground. "The goal for that project," Marcy said, "was always very clear." He would use the Keck II telescope, which
NASA had helped fund with the idea of finding planets, to
survey 166 nearby Sun-like stars, looking for radial-velocity
wobbles. "Same old same old," he said, "except we would be
looking at very high precision, and doing repeated measurements at very high cadence."

"High cadence" means they took measurements frequently,

to delineate the curve of back-and-forth motion as accurately
as possible. With Kepler, the cadence is *extremely* high—the
satellite measures the brightness of all 150,000 stars in its field
of view once every thirty minutes, and a small subset of 512
stars once a minute, the latter mostly to look for transit-timing
variations. With this project, said Marcy, which he called the
Eta-Sub-Earth Survey, the cadence would be one measurement, or sometimes two, every night, which is pretty fast for a
survey that has to go from one star to the next to the next. "The
stars were chosen blindly," he said. "We selected them without
knowledge about the planets that might or might not be around
them." Choosing stars where you know planets exist is a cheat.
It makes the survey nonrandom. Even choosing stars you think
are more likely to have planets would be a cheat—by picking
stars high in metallicity, for example, which Debra Fischer had
shown to be especially fertile planet-hunting territory.

   "We knew we wouldn't be able to find planets exactly the
mass of the Earth," he said. "Our technique can't do that, even
for the closest-in ones. You get very, very close, but you can't
find planets that are Earth mass or smaller, and certainly not
out at one AU." An AU, or astronomical unit, is the distance
Earth lies from the Sun, or about ninety-three million miles.
A Mirror Earth around a Sun-like star has one Earth mass and
orbits one AU out. In our solar system, Venus orbits at a little
over .7 AU. Mars is at a hair more than 1.5 AU. Pluto, with a
highly elliptical orbit, varies from just under 30 to nearly 50.
"So the goal," continued Marcy, "is to measure the fraction
of stars that have very small planets in close-in orbits where
our technique is very strong." There were a handful of others

doing similar projects, he said. "The Swiss team is doing the best. They're doing very good work, and they've found more than we have."

By the fall of 2010 when we spoke, the project had been under way for more than three years, and Andrew Howard had just written up the results to date in a paper that was about to come out in *Science*. "When all is said and done," said Marcy, "cutting right to the bottom line, we surveyed the planet inventory from those larger than Jupiter all the way down to the smallest we could detect, which was three Earth masses, and found that there's an ever-increasing number of planets toward lower and lower masses, down to the smallest. The funny thing about this result is that for me, this is like a lifelong dream. It was just fifteen years ago that finding a Jupiter, any old Jupiter, was amazing. Here we have the distribution of planets down to three Earth masses. It's completely unbelievable that we have come this far."

There was just one thing that worried him. "As exciting as this discovery and the new paper is, the theory of planet formation, albeit still adolescent at best, makes a distinction between the formation of rocky planets like Earth on the one hand, and the formation of Neptune and Uranus, which are mostly not made of rock, on the other." Neptune and Uranus are roughly 50 percent water; the rest is gas. "They are literally water planets," said Marcy, "with a rocky core of five to ten Earth masses." In our solar system, there's a huge gap between Neptune and Uranus, at seventeen and fifteen Earth masses (which simply means seventeen and fifteen times the mass of the Earth), and Earth, at one. Somewhere in that gap, there would presumably be a transition. Above a certain size, you form a water world. Below, you form a rocky planet. Nobody currently knows where that transition lies, but with Kepler, and more crudely, with the Eta-Sub-Earth Survey, you can start to fill in the gap and see. "We already have masses and radii from lots of planets," said Marcy, "and I can tell you that a whole lot of them down to just a few Earth masses are fluffy. They are low density. They are reminiscent of Uranus and Neptune."

"So this then goes back to our wonderful Howard et al paper," he continued, "which on the one hand is a dream come true but on the other hand offers pause for some sobering thoughts." The good news, he said, is that the number of planets is rising inexorably higher and higher as you go toward less massive planets, all the way down to three Earth masses. This is a good sign that with just a little more work the exoplaneteers will find lots and lots of planets with the mass of Earth. Presumably, said Marcy, "the Earths would have a rocky surface with continents and plate tectonics and maybe oceans and lakes—the vision that comes out of science fiction movies."

But the final verdict, he continued, was not yet in. "I'm a little bit worried," he said, "and from the Kepler data, I have reason to worry. From the theoretical side and from the observational side, we haven't answered the question yet about whether Earth-like planets are common." Planets the size of Earth, he was saying, might well not be Earth-like in any other way. He also didn't have any strong evidence to support this worry, but he said, "I have some weak evidence."

This evidence came from the theorists who try to create

virtual solar systems with computer simulations. The planets with lots of gas—Jupiter, Saturn, Uranus, Neptune—form quickly, in the first few million years, because after that the gas left over from the original collapsing interstellar cloud that formed the solar system disperses. "What's left over," said Marcy, "is dust, dust particles, maybe some pebbles of up to a half inch in size, and according to theory it takes another hundred million years roughly, maybe fifty million years, nobody knows for sure, to form the Earths." But those Earths turn out to be hard to form. The theorists let the leftover material whirl around and collide and stick together in their simulations, and, said Marcy, "you end up with Marses with no trouble, but you don't make Earths very easily." Earth, he pointed out, is nearly ten times more massive than Mars, and it's hard to gather enough material together to create one—in the simulations, anyway.

"You can't bet your house on what the theorists tell you," he said. He knew this from the experience of finding hot Jupiters, which few theorists had ever imagined. "But if this scenario is correct," he said, "seeing small planets from the Uranus and Neptune category offers virtually no information about the rocky planets. They are theoretically completely different scenarios, utterly different."

In fact, the question of how planets form, and how they should be classified, has been around since the first exoplanets were discovered. Some of the worlds Mayor and Marcy found in the first few years were six or more times as big as Jupiter. Above a certain threshold size, an object would presumably not be a planet anymore, but rather a brown dwarf—but what was the size? Some theorists suggested that brown dwarfs, which form, like stars, directly from the collapse of interstellar clouds, could actually be smaller than some planets, which grow from smaller bits of gas and dust. And then there were the Pluto wars, which ultimately forced the smallest planet to be reclassified as a dwarf planet. Part of the argument there was that Pluto was more like a huge comet than anything else.

To this day, there's no really good definition of the term *planet*, although Marcy's colleague and Natalie Batalha's mentor, Gibor Basri, did his best to come up with one in 2003. In Basri's scheme, any solar system could be divided into three kinds of objects. A *fusor* was defined as an object that achieves nuclear fusion in its core during its lifetime (this would include not only all normal stars but also the pulsars where Alex Wolszczan and Dale Frail had found the very first Earth-mass planets, since pulsars were ordinary stars before they blew up). A *planemo* was defined as a round nonfusor. Pluto, its sister world Eris, and the asteroids Ceres and Vesta, among others, are planemos. So are the Moon and some moons orbiting the outer planets (including Jupiter's Enceladus, Io, and Europa; Saturn's Titan; Neptune's Triton; and more). But these moons wouldn't classify as planets because in Basri's scheme a planet was defined as "a planemo orbiting a fusor." A planemo orbiting a planemo didn't count. Pluto, Eris, and the asteroids, on the other hand, which orbit the Sun directly, did. You've undoubtedly never heard of planemos or fusors, however, since the terms, for obvious reasons, never caught on.

"I always thought the debate over Pluto was a stupid argument," said Marcy, "and most of my colleagues thought it was a stupid argument as well. We would e-mail privately to each other that all this hullabaloo about Pluto was much ado about nothing." He was always wryly amused, however, by the fact that while everyone was arguing about whether Pluto was a planet, the eight planets everyone did agree on were, as Marcy put it, "two qualitatively different types of beasts. And no one seemed to be bothered by the fact that they are all so different and yet we call them all planets." To realize that the Earths are so qualitatively different is really important, he said, because it bears so strongly on the question of how easily and how often they form. "It's easy to pretend we have the answer to that, theoretically, but we don't."

What we do know, thanks to Marcy's Eta-Sub-Earth Survey, and Dave Charbonneau's MEarth Project, and the radial-velocity work of Mayor, and the results that were spilling from Kepler's light detectors, is that the dividing line between Neptunes and Earths might fall a lot closer to Earth than to Neptune. "We have GJ 1214 b," Marcy pointed out, "which is Dave Charbonneau's water world. And Hat-P-11 b, and Gliese 436 b, which my group found, and they all have radii smaller than four Earths, and densities of about two, which means they formed in an environment that had water and almost certainly gas as well—not at all the way the Earth formed."

If Earth-size worlds can be vastly different from the original Earth, in short, the search for life could get very complicated. This unsettling idea was circulating in the exoplanetology community well before Dave Charbonneau found the water world GJ 1214 b. In 2003, for example, a Harvard postdoc named Marc Kuchner wrote a paper for the *Astrophysical Journal*. "Discussions of extrasolar planets," it said, "often quietly assume that any object with [one Earth mass] orbiting in a star's habitable zone will be terrestrial, i.e., composed mostly of silicates and iron-peak elements like the Earth. However, we suggest that the habitable zones of nearby stars could harbor other similar-looking beasts." The beasts he had in mind in this paper were water worlds, covered with oceans many hundreds of miles deep and surrounded by atmospheres rich with steam. The French planetary scientist Alain Léger proposed a similar idea in a paper published in *Icarus* in 2004. "A new family of planets is considered," he and his colleagues wrote, "which is in between rocky terrestrial planets and gaseous giant ones: 'Ocean-Planets.'"

A few years later, Kuchner, by now a postdoc at Princeton, co-authored a paper with Sara Seager. Titled "Extrasolar Carbon Planets," it pointed out that the giant interstellar clouds of gas and dust that collapse to form solar systems aren't identical. They do have the same general mix of elements, but the proportions can surely vary. One of the likely variables is the ratio of carbon to oxygen. Our Sun has about half as much carbon as oxygen, which reflects what our original cloud was made of and what the solar nebula was made of as well. But it's easy to imagine an interstellar cloud with a different ratio. With a higher percentage of carbon, the planets that would be born of that extrasolar nebula could in principle be mostly carbon, not rock, with a core of pure diamond. Kuchner and Seager went on to ask:

What other possible kinds of planets are there? The planet
zoo now contains silicate planets (e.g., Earth and Mars),
hydrogen and helium planets (e.g., Jupiter and Saturn), water
planets [Neptune and Uranus] (which perhaps we might
think of as oxygen planets), iron planets, and carbon planets. A glance at a table of solar abundances suggests that next
we might consider helium, neon, and nitrogen planets.

At the time they were writing, of course, the zoo "con-
tained" carbon planets and iron planets in the sense that they
were theoretically possible. Seager and Kuchner had described
water planets; the iron planets they were referring to had been
described by David Stevenson, a Caltech planetary scientist.
"Yes," Stevenson said in a recent interview, "you could have a
planet that was essentially an Earth-size cannonball." Stevenson was familiar with Kuchner and Seager's work, and agreed
with the general proposition that Earth-size planets could come
in all sorts of flavors. "There's a tendency to look at our solar
system and think, 'We've got one of these and one of those' and
decide that's the whole range of possible planet types."

But our solar system lacks certain obvious types of planets.
Never mind giant cannonballs or giant lumps of carbon with
quadrillion-carat diamonds inside. "What about a super-
Ganymede?" Stevenson asked, referring to Jupiter's, and the
solar system's, largest moon. It's easy to imagine something
like that forming out of the early solar nebula; it just happened
not to. But given the enormous number of exoplanets that
have been found so far with the limited searches we've been

able to do, he said, "if you have a good physical reason to think something is possible and haven't found it yet, you probably will."

Even after Dave Charbonneau found GJ 1214 b, however, it wasn't absolutely certain that he'd found a water world. Based on its overall density, it could be half rock, half water, surrounded by an atmosphere thick with steam (since the planet is so close to its star). But it could also have a smaller rocky core with a huge thick atmosphere of hydrogen gas. The way to distinguish between the two would obviously be to take a look at the atmosphere, and fortunately, the star and planet are only forty light-years away. This doesn't make such an observation easy, but it at least makes it possible: You look at the star's light as it passes through the planet's atmosphere, taking a so-called transmission spectrum, just as Charbonneau did when he first detected sodium in the atmosphere of HD 209458 b, and see what elements or compounds are there.

If GJ 1214 b has a hydrogen-rich atmosphere, you should expect to see water. If the atmosphere is rich in water, you shouldn't. This naturally sounds a bit crazy, but Zach Berta, the MEarth team member whose "nice shooting" had spotted the planet in the first place, explained why it's not. "A hydrogen-rich atmosphere," he told me, "will still have some water in it, and because the atmosphere is physically big [hydrogen is much lighter than water vapor, so it expands to fill a bigger volume] you can see the water easily." A water-rich atmosphere, by contrast, would show very weak water features. That's not, said Berta, because there isn't much there, but because the atmosphere is dense and squashed down, it's relatively small. In early 2012, Berta finally did observations that confirmed the squashed-down, water-rich version.

There's also some evidence for carbon planets. In 2010, Princeton postdoc Nikku Madhusudhan and University of Central Florida faculty member Joe Harrington used several telescopes to look at the transmission spectrum of WASP-12b, a hot Jupiter found a year earlier by the UK-based Wide Angle Search for Planets. They found more than twice as much carbon and one hundred times as much methane (which is made of carbon and hydrogen) as you'd expect to see in a planet like this. If there's an Earth-size planet in this system, it could plausibly have a core of pure diamond, with diamond continents sloping down to seas of tar.

The very fact that Wasp-12b exists, however, might mean that its solar system has no Earths at all—and the same could be true of any solar system with a hot Jupiter. The reason it took so long to find the first exoplanet was that nobody imagined a planet as big as Jupiter hugging tightly to its star. It certainly couldn't have formed there, and the first explanation theorists could come up with was that it had somehow spiraled in from its original location, much farther out. This wasn't necessarily good news for finding Earth-like planets, since an inspiraling Jupiter would have disrupted the orbits of anything in its way, even flinging other planets out into deep space.

More recently, though, evidence has turned up to suggest a different scenario. It's been uncovered in part by Josh Winn at MIT, the astronomer-turned-science-journalist-turned-astronomer who is working on the TESS mission. In our own solar system, the planets orbit in the same direction as the Sun rotates, and in pretty much the same plane as the Sun's equator. That makes sense: The Sun and the planets all condensed out of the same original spinning pancake of gas and dust, so everything should move in the same direction.

But by using something known as the Rossiter-McLaughlin effect, which was first posited all the way back in the 1890s, Winn, Marcy, and others realized that between a quarter and half of all the hot Jupiters are way out of line. They orbit at sharp angles to their stars' equators, and in some cases even revolve in a direction opposite to the stars' rotations. "This burst on the scene about a year ago," Marcy said. "The Europeans found a few, we found a few. That was disturbing, but as a good scientist, you sweep it under the rug. Maybe it's just a fluke." But then, he said, five consecutive WASP planets were all found to be misaligned. "Now you could no longer sweep. We all smacked ourselves on the forehead."

The Rossiter-McLaughlin effect works like this: When Geoff Marcy or Michel Mayor measures a star's redshift or blueshift—the shift in light that happens when a star is moving away from you or approaching—they're actually looking at an average. The star itself may be moving away or coming toward you or standing still. But even if the star is standing still, its rotation means that one edge is moving toward you and the other is moving away.

Now, imagine a planet transiting across the face of the star. If it's orbiting in the same direction as the star, it will block the approaching, blueshifted half first, then the receding, redshifted half. So if you look carefully at the star's spectrum, you should see a subtle change in the mixture of redshifted and blueshifted light as the planet moves across. If the planet is traveling backward, though, you should see a dimming of the redshifted half first, then the blue: The change will happen in reverse. And if it's orbiting at some sharp angle, the pattern will be somewhere in between. The Rossiter-McLaughlin effect can even tell you whether a planet is traveling across the star's meaty middle or merely grazing its edge. Every possible combination of route and direction across the star's face has a unique signature.

The significance of planets that go backward and are otherwise out of kilter means that inward migration might not be the explanation for the hot Jupiters after all. "Back in 1995," said Marcy, "Doug Lin [of Santa Cruz] jumped up and said, 'I have this migration model.' It had problems, but we all kind of accepted it. It doesn't explain retrograde motion, though, which shows that our best idea for the past fifteen years was mostly wrong—or at least, wrong half the time." The best alternative explanation is that planets, as Marcy puts it, "slingshot themselves." That is, when they approach one another too closely in the early days of a solar system's existence, they tend to fling one another around with their gravity. "I find it lovely and slightly embarrassing," said Marcy, referring to the fact that exoplaneteers may have been barking up the wrong tree for a decade and a half, "but this is what makes science wonderful. Sometimes you have to throw the baby out with the bathwater."

It probably won't be the last time, either. The problem with creating a convincing theory of planet formation and motion is no longer that we have too few planets to work with. With hundreds in hand and hundreds more coming from Kepler, there are now plenty. The problem is that astronomers are still working with a very biased sample: They can see only the planets that are easiest to see. At first, these were the hot Jupiters. Now, thanks to better radial-velocity measurements and to transits, they can find smaller planets, and planets farther out from their stars.

But even with the progress we've made in planet detection, we'd still fail to find seven of the eight planets in our own solar system with all existing techniques if it were just a few tens of light-years away. "I wouldn't say it's foolish to be making theoretical predictions and working on the theory of planet formation," said Scott Tremaine, a theorist at the Institute for Advanced Study and Eric Ford's thesis adviser—the one who went off hiking while Ford was doing his research. "But I think there has been a culture in which a lot of theorists think of it as a little bit like playing the lottery. You make a prediction and if it turns out to be right, then you're a hero, and if it turns out to be wrong, it's going to kind of be forgotten about and so you don't really suffer from it."

With that in mind, here's a really intriguing theory from David Stevenson, the same Caltech planetary theorist who talks about iron cannonballs the size of Earth. Back in the late 1990s, Stevenson wrote a paper based on the slingshot idea. As Jupiter was forming—according, at least, to conventional planetary theory—smaller, Earth-size objects should have formed as well, with atmospheres made of the same hydrogen that now shrouds Jupiter in a layer of gas thousands of miles deep.

Most of these objects would have been absorbed into the giant planet, but some could have been slingshotted entirely out of the solar system. "This part is not in dispute," he said. "Jupiter and other giant planets were perfectly capable of ejecting stuff. So that's rather straightforward."

The part that is not so straightforward is Stevenson's idea that these Earth-size bodies, hurtling through interstellar space no longer tethered to any star, would retain their dense, hydrogen-rich atmospheres. Hydrogen, it turns out, is a green-house gas, just like carbon dioxide or water vapor: It keeps heat from escaping out into space. On our own planet, that heat comes from the Sun. On these lonely worlds, it would come instead from within, through the decay of radioactive elements in their cores—the same heat source that keeps Earth's core partially molten. "The physics is not in doubt," he said. "It's just a question of whether the planet would retain its atmo-sphere when ejected. I have no doubt it happens; I just don't know how often. I suspect it's fairly common."

Such a world would be pretty much impossible to detect, however. That's unfortunate, since if Stevenson is right about how common and how balmy they're likely to be, they could be ideal places for life to have taken hold. As long as we're dreaming, it's even conceivable that these lonely worlds, trapped in perpetual night, might be the most common places in the Milky Way where life has managed to thrive. We can't help thinking of Earth as the model for an inhabited planet. If this theory is correct, however, our world could be just a quirky oddball, far outside the biological mainstream.

Chapter 15

WHAT DOES "HABITABLE"
REALLY MEAN?

W HEN DAVE CHARBONNEAU originally coined the term *exoplaneteer*, it referred to someone who searches for planets around stars beyond the Sun. But Charbonneau, and Geoff Marcy, and Bill Borucki, and virtually every other planet hunter I ever spoke with was ultimately looking not simply for planets, but for life beyond Earth. That search covers a broader range of scientific disciplines than just pure astronomy, so *exoplaneteer* can just as accurately describe not just astronomers, but also planetary scientists, climate modelers, chemists, biologists, geologists, and more. Back in the 1960s, scientists who were interested in the question of extraterrestrial life (the scientific question, that is, not the question of whether UFOs were crashing in Arizona) began using a variety of names for this meta-discipline, including bioastronomy, exobiology, and astrobiology. By the late 1990s, everyone had pretty much settled on astrobiology, including NASA, which created the NASA Astrobiology Institute in 1998. It's a good, descriptive catchall term. But *exoplaneteer* is a lot more fun.

Whatever you call it, the best way to think about exoplan-etology comes from an equation scribbled on a blackboard in rural West Virginia in 1961. That's when a young radio as-tronomer named Frank Drake first wrote down the Drake Equation, a series of symbols you can use to calculate the number of technologically advanced civilizations in the Milky Way. Drake was then a scientist at the National Radio As-tronomy Observatory in Green Bank, West Virginia (the ru-ral setting was chosen because it was far from artificial radio sources that could contaminate signals from space). He was intrigued with the idea of listening for alien radio broadcasts with the observatory's powerful radio dishes—Drake per-formed the very first search for extraterrestrial intelligence, or SETI, that same year—and he'd organized a small workshop to talk about the question. A young Carl Sagan was one of the participants.

A few days before his guests arrived, Drake realized he needed some sort of structure for the meeting, so he thought for a bit, then wrote this down on the blackboard: $N = R^\star f_p n_e f_l f_i f_c L$. N—the number you are trying to figure out—represents the number of civilizations capable of communicat-ing across interstellar space. The letters on the right side represent, in order: $R^\star$, the rate at which Sun-like stars form; f_p, the fraction of stars that form planets; n_e, the number of planets per solar system hospitable to life; f_l the number of planets where life emerges; f_i, the fraction of life-bearing planets where intelligence evolves; f_c, the fraction of these planets that have developed interstellar communication; and L, the average life-time of such civilizations (if they arose and then died out quickly, there would be few of them around). If N is large, it makes sense to search for alien signals; if not, it does not.

The equation is so well known by now that people accost Drake in restaurants to have him write it down for them, along with his autograph (he once said that Japanese tourists often want him to write it on their clothing, for some reason he hasn't been able to figure out). But while it's useful as an organizing principle, nobody has a clue what N actually is. In 1961, the only term on the righthand side that anyone could put a number to was the rate of star formation. Today, thanks to Kepler and the radial-velocity searches conducted by Mayor and Marcy, exoplaneteers are on the verge of nailing down eta-sub-Earth, in one form or another.

But how often life might arise on exoplanets is still a complete mystery. At Frank Drake's 1961 conference, Carl Sagan suggested it would happen 100 percent of the time: Life should arise on every Earth-like planet in the habitable zone of a Sun-like star. It's not a crazy proposition: The basic components of life as we know it on Earth are water and complex, carbon-based molecules, both of which are plentiful in the Milky Way. Astronomers have even found such organic molecules as formaldehyde and alcohol floating in interstellar space (to a chemist, *organic* doesn't mean living, or produced by living things; it simply means *carbon-based*).

It seems plausible that water and organic chemicals must inevitably give rise to life, but that's a long way from proof. The only reason to believe such a thing is that life seems to have arisen on Earth by around 3.5 billion years ago, just a few hundred million years after the surface had cooled to tolerable temperatures in the aftermath of a bombardment by asteroids. If life arose so quickly, goes the argument, its appearance must have been pretty much inevitable. And if that's the case here, it should be the same everywhere.

Another line of reasoning that supports the "life is everywhere" theory, unknown at the time of Drake's conference, has emerged over the past few decades: Scientists have found living organisms—bacteria, mostly—thriving in an enormous range of harsh and improbable environments, including floating sea ice in the high Arctic; pools of water that are boiling hot, or harshly acidic, or salty, or even radioactive; and solid rock a mile or more underground. If life can survive in such awful places, it could easily exist on planets that are barely Earth-like at best. This could suggest Sagan's optimism may have been even more fully justified than he knew at the time.

But again, this is purely circumstantial evidence. The truth is that nobody has a clue about how life first arose on Earth, or even where. Charles Darwin suggested in passing that it might have happened in a "warm little pond." Since the 1950s, scientists have offered other ideas: It happened in the atmosphere, or in superheated water gushing from cracks in the ocean floor, or in beds of clay, or in lightning-charged clouds of gases spewing from ancient volcanoes. The best understanding of *how* it happened is similarly murky. The emergence of life must have involved a complex interplay between organic compounds that somehow organized themselves into self-replicating molecules. Here's an excerpt from the Wikipedia article on "RNA world hypothesis" that nicely captures current thinking about just one of several theories of how it all happened:

> The RNA world hypothesis proposes that life based on ribonucleic acid (RNA) pre-dates the current world of life based on deoxyribonucleic acid (DNA), RNA and proteins. RNA is able both to store genetic information, like DNA, and to catalyze chemical reactions, like an enzyme . . . It may therefore have supported pre-cellular life and been a major step in the evolution of cellular life.
>
> In a 2011 review of the evidence, Thomas Čech suggests that multiple self-replicating molecular systems probably preceded RNA . . . The RNA world hypothesis suggests that RNA in modern cells is an evolutionary remnant of the RNA world that preceded ours.

Note the words *hypothesis*, *may*, *suggests*, and *probably*. Also note that the RNA-world hypothesis isn't the only one making the rounds. There's also the "lipid world hypothesis" and the "iron-sulfur world hypothesis," and a few more. Rather than go into the details of each, let's just say that the question of how and where life arose on Earth is a massively complex puzzle. The puzzle pieces themselves—the physical evidence of what really happened—have long since vanished. The best biologists can do is to try reconstructing what the pieces might have looked like, and how they might have fitted together. Every breakthrough in origin-of-life studies to date has been an important but very small step toward a convinc-

ing explanation of how it really happened. It may be that life
is inevitable, given the right conditions, as Sagan thought. It
may equally be that life is terribly, terribly unlikely to happen,
even under the best of circumstances. The fact that life on
Earth survives in so many harsh environments, moreover,
doesn't prove that life arises easily. It proves only that that life
can adapt like crazy *after* it arises.

If you're a pessimist, therefore, you might conclude that the
search for extraterrestrial life might well prove to be fruitless.
If you need further ammunition to bolster your pessimism,
you might take a look at the book *Rare Earth*, published by
paleontologist Peter Ward and astronomer Don Brownlee in
2000. The authors advance a series of arguments to suggest
that while life might well be common in the Milky Way, the
sort of advanced life we'd really love to find is very rare. Each
argument by itself sounds pretty convincing; taken together,
they appear at first to be devastating.

Take Jupiter, for example. If our biggest planet had spiraled
in toward the Sun to become a hot Jupiter, it would probably
have disrupted Earth's orbit. But if we had no Jupiter at all,
that could be a problem as well. The reason, argue Ward and
Brownlee, is that Jupiter shields the Earth from comet impacts. Comets originate from the outer solar system, and most
of them stay there. When one does fall in toward the Sun,
however, it's almost always flung away by Jupiter before it can
get anywhere near Earth. The astronomer George Wetherill
showed decades ago that if Jupiter didn't exist, we would get
about ten thousand times more comets smashing into Earth

than we do—not a good thing for the emergence and evolution of anything more advanced than bacteria.

Ward and Brownlee also point out that our Moon is much bigger in relation to Earth than any planet-moon pair in the solar system. It's so massive that its gravity helps stabilize the tilt of the Earth. Mars, whose moons are tiny, wobbles something like a spinning top that's close to falling over. Without the Moon, our planet would do the same, making the seasons highly unstable and making it hard for plants and animals to adapt.

And then there's plate tectonics, which recycles the Earth's crust back into the interior over hundreds of millions of years. That process also recycles carbon dioxide after it binds chemically to surface rocks, ensuring that the atmosphere doesn't undergo a runaway greenhouse effect, turning our planet into a hothouse like Venus. Of all the rocky bodies in the solar system, only Earth has plate tectonics, so it's probably rare in the universe. And then there's Earth's magnetic field, which protects us against energetic particles streaming in from the Sun or from deep space. And then . . . well, suffice it to say that *Rare Earth* makes a sobering read.

It does, that is, until you talk to Jim Kasting. "A lot of people read [*Rare Earth*] and believed it," he told me during our conversation at that Vietnamese restaurant in Seattle. "I think they sold a lot of copies because it was the anti–Carl Sagan. It appealed to people who didn't want to believe this whole line of stuff that Carl had been selling."

One by one, Kasting addressed the arguments in *Rare Earth*

and made it clear that he wasn't impressed. For example, he said, it's true that if you eliminated the Moon, Earth's tilt would wobble chaotically. But if Earth were spinning faster—if the day were twelve hours long rather than twenty-four—the chaos would go away. "So you have to ask," said Kasting, "How fast would the Earth be spinning if you didn't have the Moon? And that's complicated." In short, Ward and Brownlee raise a plausible argument, but hardly a definitive one.

It's also true, continued Kasting, that Jupiter protects Earth from comet impacts. But it actually *raises* the odds we'll be struck by asteroids. That's because the asteroid belt is just Sun-ward of Jupiter, so it's relatively easy for the giant planet to nudge a mountain-size chunk of rock into an Earth-crossing orbit. "It appears," Kasting writes in his 2010 book *How to Find a Habitable Planet*, where he devotes a full chapter to pre-senting counterarguments to *Rare Earth*, "that having a Jupiter-sized planet . . . is a mixed blessing."

As for plate tectonics, he said, Venus is the only other planet in our solar system besides Earth big enough to have them in the first place (a planet smaller than Venus would have cooled off by now, so it wouldn't have the semi-molten rock that al-lows continents to slide around). But Venus lacks the water it would need to lubricate the motion of crustal plates, which could be why, despite its adequate size, it doesn't have plate tectonics. Out of two planets that might have plate tectonics, one of them does, and Kasting sees no reason at all to assume that Venus is somehow typical of exoplanets while Earth isn't. The bottom line, he said, is that "there are a lot of things that we don't know, so we make conjectures. Ultimately, if we can do TPF and follow up with post-TPF missions, we'll figure out what happens, and where." "I'm an optimist," he admitted. "I agree with Carl Sagan. I think there's probably life all over the place, and there are probably other intelligent beings. I'm just not as good at speculating as he was."

There's another reason you might lean in the direction of optimism. The concept of the habitable zone applies if you're assuming life is confined to the surface of a planet. If you discard that assumption and consider places where conditions are favorable beneath the surface, you've suddenly got a lot more places to look. In our own solar system, Earth has the only habitable surface, but planetary scientists think the Martian subsurface might be habitable as well. In November 2011, NASA launched its biggest, most capable rover toward Mars, where the six-wheeled, SUV-size *Curiosity* will, among other things, drill into the Martian soil to look for organic chemicals (but not, on this mission, for life itself).

The right conditions for life could also exist on even more exotic worlds. Astronomers have known for years that Jupiter's moon Europa and Saturn's moon Enceladus both have subsurface water. The energy to keep the former from freezing solid right down to the core comes from tidal squeezing, as it orbits through the powerful gravitational field of Jupiter; Enceladus's heat source is a mystery. More recently, theorists have suggested that even Pluto might harbor liquid water, one hundred miles or so beneath its icy surface—the heat in this case coming from the decay of radioactive potassium. As for complex carbon molecules, they're abundant in the bodies of both comets and asteroids, which have been crashing into the moons and the outer planets for billions of years.

Yet another plausible reason for optimism arises from the fact that the universe is under no obligation to follow the "life as we know it" rule. Carbon is abundant in the Milky Way and combines easily with other atoms to form the elaborate organic molecules that underlie all of terrestrial biology. Water is abundant as well, and acts as a versatile solvent. So it's not absurd to think that carbon-based life might be universal, and is exactly what astrobiologists should be looking for. "It may turn out to be universal," Dimitar Sasselov, Sara Seager's grad school thesis adviser, said on a visit to his Harvard office. "There may be some basic law of chemistry, which always leads you to the use of amino acids and nucleic acids for coding and the use of particular metabolic cycles for energy." "But," he added, "it also may be environmentally dependent."

Sasselov now directs Harvard's Origins of Life Initiative, an interdisciplinary effort to understand where life comes from, and under what conditions, in order to guide future observations. This sort of astrobiology collaboration, in which biologists and geologists and astronomers and planetary scientists try to work together, is very popular nowadays; Debra Fischer has a similar collaboration at Yale, for example.

At Harvard, Sasselov's group is thinking about alternate biologies—particularly, he said, on planets where the global geochemical cycle is based not on carbon but on sulfur. "We showed in a paper last year," he said, "that a sulfur cycle on a nearby Earth or super-Earth would be easily detectable with the James Webb Space Telescope. Not only is it detectable, but you'll be able to measure the relative concentrations of sulfur dioxide and carbon dioxide and water, which is what the chemists in our group need in order to set up the experiments." In the meantime, the biochemists in the group are doing a sort of practice run, trying to create an alternate biology in the lab that's still based on carbon, but whose DNA and amino acids twist in the opposite direction from those in all Earthly organisms. As far as anyone knows, this mirror life would violate no rules of biochemistry, and while Earth biology is based on "left-handed" amino acids, their mirror-image right-handed counterparts also exist in nature.

The goal, explained Sasselov, is to create a primitive living cell. "That's why we're doing the mirror project. It's trivial from a planetary point of view," he said, since it's still a form of carbon-based life, "but it's the easiest way to develop the basic methodology, which we'll then use for a more weird biochemistry—weird in the sense of an alternative system." The project, he said, is "pretty well along. George [Church, a geneticist and biochemist at Harvard Medical School] thinks that we're within months of finishing it, and Jack [Szostak, ditto] thinks maybe a year and a half." "What we expect from the experiment is a chemically functioning system," he added, "not something that's going to walk on this table. But that's all we need."

It hasn't escaped Sasselov that making a functioning cell with a biochemistry found nowhere on Earth sounds like science fiction. "That's why I am attracted to this field," he said. "I see a direct connection between what I do and the big questions. It would be exciting to say, 'Well, I managed to detect water and sulfur dioxide on that exoplanet,' but it's not exactly one of the big questions of science. 'What is the nature of life?' is a big question."

The other big question that Sasselov is trying to answer, in his role as a Kepler co-investigator, is how big a planet can be and still be habitable. If a world has to be a true Mirror Earth in size, there will obviously be a relatively small number of planets to choose from: You're going back down the road of pessimism. Sasselov isn't going down that road. "If you're asking what the optimal size is for life, I don't see a dividing line," he said, "between one Earth mass and five Earth masses. In fact, if you ask me, bigger is better. Smaller is not. Mars is definitely too small. If you go much below an Earth mass, you can't have plate tectonics, and you don't have a stable atmosphere because it evaporates too easily [because there's less gravity to hold onto it]." Earth is not a Goldilocks planet from this point of view, he said. It's not "just right" for life: It's at the small end of the habitable range.

All these factors—how likely life is to arise in the first place, what environments allow it to arise most easily, how many planets there are, of what size, in what orbits, with what geology and geochemistry, with what other sorts of planets in the same system—feed into the question of whether life exists anywhere but Earth, and if so, whether that life is more advanced than a bacterium. Nobody really knows the answer to any of these questions. Nobody can give even a ballpark solution to the Drake Equation, a half century after it was first written down. Nobody knows what *habitable* really means.

When the question comes up at conferences, as it frequently does, this lack of knowledge doesn't keep exoplaneteers from weighing in with their own best guesses, of course. At one meeting, however, I heard a different response from Dave Charbonneau. How do you define *habitable*? someone asked. "I don't much care," he said. "I want to build an experiment that can find things as small as the Earth and that are roughly at the same irradiance taking into account the luminosity of the stars, and then you can ask me in ten years what makes a planet habitable. We probably won't know but I think that's the way that we'll make the greatest progress."

It's also possible that the small handful of astronomers who are still doing SETI searches will detect an alien signal long before Dave Charbonneau's ten years are up. This could, in principle, happen tomorrow. It's been a half century since Frank Drake wrote down the Drake Equation and in all that time not a single verified transmission has been picked up. That hardly means the search has been a failure, however, argues Jill Tarter, director of the Center for SETI Research at the SETI Institute. The search has suffered from limited resources from the start, she says. "If you dipped a drinking glass into the ocean once," she likes to ask, "and came up without a fish, would you conclude that there are no fish in the sea?"

Our failure to hear transmissions, in other words, doesn't mean they're not out there. The men and women who have continued to listen for alien radio signals for the past fifty years and, more recently, for possible flashes of light from alien signaling lasers, have always been mindful of Philip Morrison and Giuseppe Cocconi's observation that the probability of success is difficult to estimate. Even if millions of habitable planets exist, which is looking more likely all the time, and even if some of them have given rise to technologically advanced life, nobody really knows whether radio waves or laser beams are the standard way of talking across interstellar space. Maybe what we think of as advanced communications is their equivalent of smoke signals—a stepping stone on the way to technologies we can't yet imagine.

Chapter 16

A WORLD MADE OF ROCK,
AT LAST

BY THE TIME the American Astronomical Society's winter meeting rolled around in January 2011, the astronomers who had gathered in Seattle were itching to get their hands on the four hundred planet candidates the Kepler team had dangled, then snatched away, six months earlier. It wasn't quite time, though. Bill Borucki had promised the four hundred would be set free on February 1, and he wasn't going to move the date forward.

The conference-goers wouldn't have to leave empty-handed, however. In what was turning out to be a pattern, the Kepler team wouldn't release candidates, but they would announce actual, confirmed planets, just as they had at the Washington meeting a year earlier. In Washington, the big news was simply that Kepler could find planets, and that radial-velocity measurements could figure out the masses and densities for some of them. The only announcement in the year since then had been Kepler-9, the first planetary system where the mass of the planets had been measured, not with radial velocities, but with transit timing, the powerful new technique nobody had even thought about when the Kepler Mission had first been approved.

This time, Natalie Batalha would be making the big reveal, at a press conference on the first day of the meeting. She agreed to brief me the day before, however, as long as I promised, on pain of her extreme disapproval, that I wouldn't say anything about it before the official presentation. We met at a hotel a few blocks from the main conference hotel; the Kepler team was having a preconference workshop to talk about the upcoming data release the next month and to work through various routine issues that inevitably came up in analyzing and following up what the satellite was telling them. When I showed up, an Ames press officer went into the workshop to get Batalha. I could see a couple dozen astronomers, including Dave Latham, Dave Charbonneau, and Bill Borucki, listening to someone making a presentation. I asked if I could just sit in on the meeting. The press officer and Batalha both looked startled for a moment, then realized I couldn't be serious. We all knew there was no way I was getting inside that room.

"So what we're announcing tomorrow," she said, when we sat down, "is the discovery of our first rocky planet, Kepler-10b. It's in a very short period of less than a day, which makes it very similar to CoRoT-7b, if you're familiar with that result." What made this discovery special, though, she said, "is that our error bars on all of our measurements are very tight. We know unquestionably that this is a rocky world." An error bar is another term for uncertainty, the plus-or-minus that goes with astronomical measurements of pretty much anything—

the age of the universe, the distance between the Earth and the Sun, whatever. "And the reason that we know," she said, "is because we've got all of our best capabilities coming together for this one discovery. We've got this exquisite photometry [the change in brightness as the planet moves in front of its star]. We've got a very high-precision radial velocity. And this star is bright enough that we can use astroseismology, which allows us to derive the fundamental stellar properties to within 2 to 6 percent accuracy."

The team had known about this planet, or, at least, had known it was a good candidate, almost as soon as Kepler was launched. "We actually saw it during the commissioning phase," she said, "before science operations had formally started." Geoff Marcy's group did its first radial-velocity measurements from the Keck in August 2009, about the time Marcy was willing to admit publicly only that the satellite was working properly. "So the mass of this planet is 4.65 Earth masses," Batalha continued, "and the radius is 1.4 times that of the Earth." The density, she said, works out to 8.8 grams per cubic centimeter, which makes Kepler-10b half again as dense as Earth. "This seems kind of high," she said. "If you google it, you'll find it's about the density of an iron dumbbell."

But that doesn't mean it's made of iron. Kepler-10b is so massive that it crushes itself down under its own weight. If you take the material the Earth is made out of—the same proportions of silicate rocks and iron and nickel—and just add more of them, the planet grows heavier at a faster rate than it grows larger, thanks to the crushing; and at 4.5 times the mass of the real Earth it would in fact be about as dense as a dumbbell. A dumbbell that heavy would be even denser. If you graph the size versus the mass of an Earth-type planet, and locate Kepler-10b on the graph, she said, "the error bar almost kisses that line of Earth composition. It's just a little higher in density, so it does seem to be a little bit iron-rich. Kind of like Mercury." CoRoT-7b kisses the Earth-composition line too, she said, echoing what Marcy had told me earlier. The difference is that, unlike Kepler-10b, it has huge error bars. "We've got three different papers on CoRoT-7b with mass estimates that range from, like, zero to ten. Zero is a funny way of saying it, but it's basically a nondetection all the way up to ten Earth masses."

For that reason, she was arguing, Kepler-10b, not CoRoT-7b, should be counted as the first unambiguously rocky world ever found beyond our solar system. It was too big and far too hot to be a true Mirror Earth, but it was another step closer. The star itself, Kepler-10, is about 560 light-years from Earth, which got Batalha thinking. "I told this story to the science team this morning," she said. "Just out of curiosity, I subtracted 560 from the year 2010 and got 1450. The year 1450 is when the light left the star. And I googled the year 1450 to see what was happening—in Wikipedia it's listed as being the beginning of the age of discovery. Isn't that amazing? I love that. So Europeans were crossing the Atlantic for the first time when light left the star. I thought that was kind of a nice tie-in." (It was now 2011, but since the distance to Kepler-10 isn't known down to the light-year, it's a legitimate bit of scientific poetry.)

A videographer named Dana Berry, who has done work for

National Geographic, had been commissioned to create a video animation of what Kepler-10b might look like—a rocky planet, hugging its star with one face always turned to the searing light glowing red with heat. "When I saw it," Batalha told me, "I thought about our small telescope up at the Lick Observatory." This was the Vulcan telescope, which Bill Borucki used to prove to NASA that he could detect multiple transits, named for the planet astronomers once thought orbited Sunward of Mercury. "So when I saw the animation," said Batalha, "the first thought that came to my mind was, 'Wow, this is our planet Vulcan.'"

Whether it might ever be named Vulcan in a formal sense is unclear. So far, the International Astronomical Union, whose job it is to assign official names to heavenly objects as opposed to the numbers and letters different catalogs use to list them, has not come up with any sort of exoplanet-naming convention. According to Geoff Marcy, his wife had an idea back in the 1990s, when he started finding his first planets. "She said I should call them Susan 1, Susan 2, and so on," he told me. Although it should be obvious that she was kidding, it's probably a good idea to make it explicit: She was kidding. When 51 Peg b was discovered by Michel Mayor, a suggestion was floated that the planet be named Bellerophon, the name of the mythological Greek hero who tamed the flying horse Pegasus. One of the pulsar planets was unofficially named Methuselah (the planet was probably very old), and HD 209458 b, the first transiting planet, was called Osiris, after the Egyptian god, but these, too, were never considered official. I've rarely if ever heard them used by astronomers.

When the press conference rolled around the next morning, one more thought had evidently come to Batalha's mind.
Or more likely, it wasn't a thought, but something subcon-
scious that led her to speak of the planet in an unusual way.
"Today," she said to a group of reporters, anxious for the latest
Kepler news, "we're announcing a new planet. She orbits her
star . . ." Throughout her talk, Batalha referred to Kepler-10b
as "she." When it came time for questions, my hand shot up.
"How," I asked her, "did you determine the gender of the
planet?" The room broke up and Batalha looked a little embarrassed. I was worried that I'd stepped over the line, even
though I was obviously teasing, but Geoff Marcy, who had
been up on the podium with her to provide commentary, as-
sured me that I hadn't. "That was a great question!" he said to
me after the press conference was over. "I was surprised because
she was so consistent. I was like, okay, I heard it once, now I just
heard it again . . . I don't know how much of a gender issue
we still have in astronomy, since there are a lot of women at
this meeting. But then," he said, after thinking for a moment,
"astronomy faculties are still mostly male, and the Kepler team
is more male than female by a large margin. So the issue really
hasn't gone away, and I think the men and women of Natalie's
generation are still rightly sensitive to the issue."

Kepler-10b was the closest thing to a Mirror Earth the Kepler team or anyone else had yet found, but it was only one of
the four hundred planet candidates that had been held back the
previous June. Just about three weeks after the Seattle meeting,
the candidates were released at a press conference at NASA
headquarters. When Bill Borucki began speaking before a

crowd of reporters, NASA officials, and TV cameras, however, he didn't just announce the four hundred; in the intervening half year, the team had vetted hundreds more. When
the press conference was over, the number of planet candidates
had leaped from 706 to more than 1,200. They broke down
this way: 19 were larger than Jupiter; 165 were about the size
of Jupiter; there were 662 Neptunes; 288 super-Earths; and,
finally, 68 planets about the size of Earth, a few of which were
a bit smaller than Earth itself. Slice it a different way, and you
had 54 planets in their stars' habitable zones, 5 of which were
approximately Earth-size (all the members of this last group
were orbiting M-dwarfs, whose habitable zones were very
close to the stars and where you could see the required minimum of three transits in just a few months).

This was the piñata Dennis Overbye was talking about in
the *New York Times*, a harvest of planets so overwhelming
nobody knew what to do with them all. If Kepler had been
looking across the entire sky rather than at a tiny patch, Borucki
said, it probably would have found about four hundred thousand planets. That's only in the nearby sky, not the Milky Way:
Kepler is a powerful telescope, but not powerful enough to
find planets much more than a few hundred light-years away.
The Milky Way is about a thousand times bigger than that.

Kepler had found *potential* planets, Borucki emphasized;
most of them would never be fully confirmed. Nevertheless,
statistics based on the planets that had been confirmed suggested that about 90 percent of these objects were real—and
these confirmed planets had come from only the first six months
of Kepler observations. Having so many candidates brought the

Kepler scientists a long way toward their goal of calculating the frequency of Mirror Earths. The team would continue to try confirming as many planets as it could, of course, with every trick it could come up with. Kepler's primary goal was statistical, but finding real planets that could be studied, either with existing telescopes or with the more powerful telescopes that might be coming over the next decade, was a close second. The day after Bill Borucki's press conference, in fact, the cover of *Nature* featured an artist's rendering of the Kepler-11 system—the next one to be confirmed after Kepler-10b, Natalie Batalha's hot, rocky Vulcan. This was the six-planet system for which Daniel Fabrycky had created a simulation that so fascinated his children—five planets ranging in size from super-Earth all the way up to near Neptune size, all huddled up against their star. If you took five planets bigger than Earth and crammed them inside the orbit of Mercury, you'd have Kepler-11.

"I was the one who first saw this system in the data and said, 'This is really interesting, we should put a lot of effort into it,'" Jack Lissauer, the Kepler scientist who served as lead author on the *Nature* paper, told me the day before it was formally released. That effort included a campaign to monitor the star itself very carefully, to be absolutely sure about its physical size, and to make calculations, based on transit-timing variations, to confirm the planets and get their masses—the job of Fabrycky, the paper's second author. (There were thirty-nine authors in all, including Eric Ford at number three; Bill Borucki, fourth; Geoff Marcy, sixth; Natalie Batalha, eleventh; Dave Charbonneau, sixteenth; Dave Latham, thirtieth; and Dimitar Sasselov, thirty-seventh. It was a who's who of exoplanetology.)

What made the system so intriguing was partly the question of how it came to exist in the first place. All of the planets were presumably born farther out and migrated or were flung into their present positions. Packing them in so tightly, however, wasn't easy for theoretical models to do. Another surprise was that the planets all orbited in precisely the same plane. "This was also going to be tough to explain," said Lissauer. "Our solar system is *approximately* flat, but given the chaotic process that gives rise to planetary systems, it's hard to see how this one came together. It's so unexpected, and we're getting so much information from this system," he said. "I think this is the biggest thing in exoplanets since the discovery of 51 Peg b in 1995," he continued. "My view is that a true Earth-mass, Earth-size planet in the habitable zone is the only thing that could be more interesting than this system. But until we find that, they'll all just be hot rocks."

So far, the Kepler team had been talking about two categories of exoplanets. There were candidates, more than 1,200 of them as of early 2011. And there were confirmed planets—about 20, distributed around seven stars, and confirmed by either a radial-velocity signal or evidence from transit-timing variations. In May 2011, at the smaller, spring meeting of the American Astronomical Society, they began talking about another category: validated planets. These weren't quite confirmed, but had passed enough tests to rule out false positives that you could consider them planets anyway. The chances were just too small that they could be anything else.

One way to validate a Kepler planet was to observe a transit in infrared light, with the Spitzer Space Telescope—Dave Charbonneau's project under the Participating Scientist Program. The dip in light from an actual transit should look the same with Spitzer as it did with Kepler; an eclipsing binary should show a color shift. Another newer and more rigorous technique, however, didn't involve any observations at all. Instead, it relied on a computer-simulation technique called Blender, originally created by Harvard astrophysicist Guillermo Torres to rule out false positives in the OGLE survey. This was the ground-based survey originally designed to look for chunks of dark matter in the Milky Way, then redirected to look for the Einsteinian microlensing, or magnification, effect of distant solar systems on even more distant stars, then redirected again to look for transits.

As Bill Borucki had realized early on, a dip in starlight could be the signal of a planet, or it could be coming from a pair of eclipsing binary stars peeping over the shoulder of the target star—a blend of light from several stars, mimicking a transit. "OGLE," Torres said at the meeting, which was in Boston, "looks at a very crowded field of stars, where the likelihood of a blend is much greater than with other surveys." So he came up with Blender, which took all possible scenarios—a wide range of combinations of foreground and background stars of all brightnesses and sizes and relative distances from Earth—and used them to calculate how likely it was that the signal the telescope saw represented one of these combinations.

"When Kepler came along," Torres said, "I realized that

they wouldn't be able to confirm most of their candidates." He approached the team and proposed they use his simulation. "They were very enthusiastic," he said, "and it's been very successful." The first success was announced by his Harvard colleague Francois Fressin at the meeting. Using Blender, Fressin and Torres were able to validate the fact that Kepler-10b, Natalie Batalha's hot, rocky Vulcan, has a companion, 10c, just a little bigger, at about 2.2 times the radius of Earth, with a forty-five-day orbit. With Blender, said Torres, "we're fairly confident that what we're looking at is a planet." By fairly confident, he explained, he meant that "we're looking for odds ratios of something like a thousand to one. That's when we call it a validated planet. It's not confirmed: There's a difference in the terminology here."

Torres and his colleagues also used Blender to validate the sixth planet in the tightly packed, CD-flat Kepler-11 system. "We're now working on other candidates," he said, "which have already been submitted for publication." And they've used Blender to validate the existence of CoRoT-7b, the possibly rocky planet that was possibly found before Kepler-10b—although Torres, like everyone before him, couldn't nail down its mass. Could Blender turn out to be the only way to validate a true Mirror Earth in Kepler's list of candidates?

His straightforward answer: "Yes."

Chapter 17

ASTRONOMERS IN PARADISE

I F   Y O U   A S K me, it's best to arrive at the Jackson Lake
Lodge, about thirty miles north of Jackson, Wyoming, at
night. The lodge is just inside Grand Teton National Park, so
you know there must be some pretty good scenery out there
somewhere, but in the dark it's purely hypothetical—and unless there's a full Moon out, it's really, really dark. When you
check in at the front desk, it feels like an ordinary hotel, aside
from the western-themed artwork and the decorative wagon
wheels and antlers sprinkled around the lobby. Most of the
guest rooms aren't in the main building, but rather in a series
of cottages fanning out along small, poorly lighted roads. If
you haven't got a car, they hand you a flashlight so you won't
get lost. The bears, the clerks promise, are very unlikely to eat
you. Most of the cottages are nestled in stands of pine trees, so
even in full daylight, as you walk back to the lodge for break-
fast the next morning, your sense of nature may be limited to
the clean air and woodsy smell.

But when you reenter the main lodge, walk past the front

desk, and go up a wide set of stairs, everything changes. You find yourself in an enormous room, probably one hundred feet long, eighty feet wide, and three stories tall, filled with couches and comfortable armchairs gathered in clusters. In the corners to your left and right are two enormous circular fireplaces, each about ten feet across. A little bit farther into the room, on the right, is a stuffed grizzly bear inside a glass case, rearing up on its hind legs with its front paws brandishing sharp, three-inch claws at just about the level of your eyes. Or, no, it isn't a grizzly after all. When you read the plaque next to the case, it turns out to be a brown bear, the grizzly's close cousin. Real grizzlies do roam the woods and grasslands around here, as do ordinary black bears. Brown bears don't: This monster was shot in Alaska. The ecological incongruity doesn't seem to matter to the tourists: They line up to pose with the bear. He towers over the tallest of them.

Imposing as they would be in any ordinary setting, though, you don't notice either the bear or the giant fireplaces at first. That's because right front of you, the western end of the room is a wall of windows framing a view so spectacular that it produces something like a physical shock. In the foreground, there's a broad expanse of flat, brushy, marshy land where you can sometimes see moose, elk, and the occasional grizzly. Beyond that is Jackson Lake itself, fifteen miles long and seven miles wide and filled with at least three species of trout along with countless other types of fish. And towering over it all, right in the center of the window, is rocky Mt. Moran, more than twelve thousand feet high, flanked by a series of less lofty peaks of the Teton Range (Grand Teton, the highest mountain in the chain, is five or six miles to the southwest, but you can't see it from this perspective).

Over the course of my five-day visit, the mountain would take on all sorts of dramatically different looks, now glowing a warm reddish gold as it was lit by the rising Sun; now silhouetted against the slowly darkening western sky after sunset; now shrouded at the top by a violent thunderstorm, or at the bottom by a thick bank of fog hovering above the lake. Each of these, and a dozen more, was worthy of a postcard (and something like each, undoubtedly, appears on an actual postcard in one of the several gift shops inside the lodge).

The view wasn't lost on the several hundred astronomers who had come here in September 2011. They would frequently stop and gaze out the windows, or step outside to take it in without the intervening glass. Many of the scientists had tacked on extra days for hiking or rafting; several had brought their families. But while the view and the outdoor activities were the only reason most visitors were here, it was secondary to the astronomers, who spent most of their days in a shade-darkened meeting room watching one presentation after another on the latest discoveries in the search for planets around other stars. The conference was titled Extreme Solar Systems II; it was a sequel to the first Extreme Solar Systems conference held in Santorini, Greece, in June 2007. That first one was convened in honor of the discovery of pulsar planets fifteen years earlier—the first solar system ever found beyond our own, and an extreme set of worlds by any definition.

That was just the excuse for a conference, however; most of

the talks in Santorini, like most of the talks here, had been about the more conventional worlds discovered since then. Many of the world's leading exoplaneteers were here, mingling unobtrusively with the tourists. (I ran into someone I knew from home, unexpectedly. "What are you doing here?" she asked. "What's all this about 'extreme solar systems'?") Looking around during the morning or afternoon coffee breaks you could see Geoff Marcy, Michel Mayor, Dave Charbonneau, Natalie Batalha, Bill Borucki, Dave Latham, Eric Ford, Jack Lissauer, and many more eminent observers and theorists, deep in conversation with one another and with the newest generation of exoplaneteering graduate students and postdocs.

Even back in 2007, there had been plenty to talk about, of course. It had been twelve years, at that point, since Michel Mayor had announced the discovery of 51 Pegasi b. Between them, Marcy and Mayor had found hundreds of planets since then, and the astronomers who had flooded into the field in the aftermath of the great revolution of 1995 and 1996 had found dozens more with their own searches. They'd found exoplanets using radial-velocity wobbles and transits and gravitational microlensing. But in 2007, Kepler was still nearly two years away from launch.

Now, everyone agreed, Kepler had changed everything. The 1,235 candidates that had been released in February had completely overwhelmed the community's ability to keep up. Even now, more than a year and a half after Bill Borucki had announced Kepler's first five confirmed planets, only 20 or so of the 1,235 candidates had crossed over the line to qualify as actual planets—although, as Geoff Marcy told me during one of the coffee breaks, "we know that most of them are going to be confirmed." But the number 1,235 was about to go out of date. There were already more candidates in the pipeline, a fact that was evident in an exchange on Natalie Batalha's Facebook page, where she and Geoff Marcy got a little bit silly sometime in the small hours of the morning just before the conference got under way.

BATALHA: Savoring the moment just before calculating the number of new planet candidates by looking out at the Grand Teton wilderness. Best. Job. Ever . . . Let the history books show that numerous new planet candidates were vetted with the aid of Cocoa Crunchies in the middle of the night.

MARCY: Wait! Surely you mean "Cocoa Crunchy Candidates." A Cocoa Crunchy ain't nothin' until vetted by difference munching and high resolution smelling. You should take the Fourier Transform of each and every Cocoa Crunchy. Admittedly, they never taste quite as good when you transform them back.

BATALHA: @geoff: severe scattering, tidal disruption, and accretion have annihilated all cocoa crunchies. No candidates left. Abort mission! ;)

The dozens of talks that would happen over the four days of Extreme Solar Systems II (with Wednesday afternoon left clear for the exoplaneteers to take hiking excursions into the surrounding wilderness) covered every conceivable aspect of exoplanetology. "It's really quite wonderful," Bill Borucki said during a coffee break after a session on planet formation. "I've just been taking a lot of notes. There's just so much here."

I knew what he meant; I mentioned to Geoff Marcy during another break that I felt like I was falling farther behind every minute. "Welcome to the club," he said. "What do you think we feel like?"

Borucki knew beforehand about some of the announcements that would be made, of course: No exoplanet conference could take place these days without a whole series of new results from Kepler, and Borucki was naturally up to date on all of these. One had actually come out just a week before the conference began: Sarah Ballard, a Harvard graduate student, had been the lead author on a paper for the *Astrophysical Journal* announcing the discovery of Kepler-19b and -19c, orbiting a star about 650 light-years from Earth. The first planet, 19b, was found straightforwardly: It transited across the face of its star.

It hadn't been confirmed, since the star, Kepler-19, was too far away and too faint to make it possible to pick up radial-velocity wobbles. It was validated, though, by Dave Charbonneau's Spitzer Space Telescope follow-up program, which confirmed the transit in infrared light. And thanks to variations in the timing of 19b's transit, Ballard's co-author Dan Fabrycky, the creator of the Kepler Orrery, was able to identify a second, unseen planet. It was something like Matt Holman's inference of a third, unseen planet in the Kepler-9 system a year earlier, but this one was more solid.

An even bigger announcement would come out toward the end of the meeting. The Kepler exoplaneteers were concentrating their attention mostly on single stars, even though there are plenty of double stars in the telescope's field of view. The reason is that for years, theorists had considered it something of a long shot that a planet could form and survive in a double-star system, where the ever-changing gravitational pull of two different suns would tend to make for an unstable environment. "When two elephants are waltzing," I wrote in *Time*, "it can be difficult for mice to tiptoe safely under their feet."

Still, Kepler sees all, and since its launch the satellite has discovered plenty of binary stars, including two thousand that eclipse each other. One such pair, which orbits each other every 41 days, showed just the sort of pattern of dimming you'd expect as the light of two stars becomes the light of one, over and over. A team led by the SETI Institute's Laurance Doyle noticed that there was a tiny bit of extra dimming, however, which showed up every 229 days, on average. There was clearly a planet here. It was the size of Saturn. Its transits varied by about 9 days in either direction, because the stars themselves were in different positions each time the planet came around. A transit would start earlier when the stars were side by side and later when they were one on top of the other— and in the former case, the transit would last longer as well, as it had to cross two stars rather than one.

Not only that: Because of the complex interplay between the gravitational tugs of the stars on each other, and on the planet, and of the planet on the stars, the transits and the eclipses all showed the sort of transit-timing variations Matt Holman and

Sarah Ballard had found in the other Kepler systems. By solving this cosmic three-body problem, Doyle and his co-authors could nail down the masses of the stars and the planet with terrific accuracy. Once they knew how massive the stars were, they could calculate their physical sizes, so the size of the planet was easy to determine with high accuracy as well.

All of this might have been a little technical to get the general public as excited as the astronomers clearly were, except for one obvious but brilliant marketing move by the NASA press office. In the first *Star Wars* movie back in the 1970s, Luke Skywalker's home planet, Tatooine, orbited a pair of suns. In one eerie scene, you see the two suns sinking slowly toward the horizon together at the end of the day. So the NASA people put *Star Wars* right in the first paragraph of their press release, and brought in John Knoll, a special-effects supervisor for George Lucas's Industrial Light & Magic, to sit on the expert panel at the press conference (they tried to get Lucas himself, evidently, but he was unavailable). The director had been well aware, Knoll revealed, that theorists didn't think such a planet was likely to exist. He decided to go with it anyway. NASA's strategy worked: The discovery generated headlines around the world. It also became clear just how many exoplaneteers are *Star Wars* fans: They threw around the name Tatooine as though it needed no explanation—which for me, it actually did.

While the Tatooine discovery was the most accessible to the public, the talk Natalie Batalha gave on the first day of the meeting was also important. In it, she announced the planet candidates she had calculated during her Cocoa Crunch–fueled all-nighter. There were now 123 Earths on the list, up from the 68 Bill Borucki had announced the previous February; 412 super-Earths, up from 288; 988 Neptunes, up from 662; and 204 Jupiters, up from 165. "It's about a 45 percent increase in candidates overall from the February release," she said during one of the lunch breaks. The statistics available to calculate eta-sub-Earth were even richer than they had been a few months earlier—partly due to the longer observing time, but partly due also to a new algorithm in the software pipeline that screened out variability in the stars themselves more effectively.

Kepler's wasn't the only exoplanet census that was featured on the first day of the conference. Michel Mayor also stood up before the audience to announce no fewer than fifty new planets of his own, including sixteen super-Earths—an extraordinary haul for any search other than Kepler. Among them was a world named HD 85512 b, just three and a half times as massive as Earth, orbiting in at the edge of its star's habitable zone (no word on whether it might be rocky, since the Swiss team didn't have a transit to calculate its density). The Swiss had also analyzed all the planets they'd found to date by size and orbital distance and come up with the claim that 40 percent of Sun-like stars have at least one planet smaller than Saturn.

Geoff Marcy, Andrew Howard, and the American Eta-Sub-Earth Survey team, meanwhile, had come up with their own figure of 15 percent, but that wasn't a direct comparison, since the Swiss, with their more sensitive HARPS spectrometer, included planets out to a larger orbital distance. And then there was Kepler, which didn't address mass at all (except in a few cases) but rather size—which might or might not be directly translatable.

The result, said Marcy, was that "by the end of Monday the meeting was left in an unstable mode. Natalie got up to discuss five hundred new planets—or planet candidates, but we all know they're mostly real. Andrew Howard got up to discuss the statistics of the occurrence rates of planets from Kepler and from radial velocity, looking at a different set of masses and different orbital distances. And Michel got up and did similar things." But without a good way to compare the three sets of data, it was hard to tell whether they agreed or disagreed. "One speaker," said Marcy, "who I won't name, suggested that there was a factor of ten disagreement. But there actually isn't."

He felt confident in saying so because, said Marcy, "in the intervening seventy-two hours we've been working behind the scenes, especially Andrew Howard. And it turns out the occurrence rates of the planets of different masses, different sizes, and different orbital distances agree. Kepler, Eta-Sub-Earth with Keck, and the Swiss team, they all agree." This still wasn't about true Mirror Earths, or even Earth-size planets. The radial-velocity teams hadn't measured such subtle motions, so only Kepler had any data at all about worlds this small. But Kepler, the new kid on the block, was getting the same answer for hot Jupiters that Marcy and Mayor were getting.

(While nobody has yet found an Earth-mass planet with radial-velocity wobbles, Michel Mayor told me at the meeting that his team was now capable of doing so. "Yes, yes, yes, yes," he said, when I asked him to confirm what I thought I'd heard him say during his talk—that he could not only detect an Earth-mass planet, but could do so in an Earth-like orbit. He could find a Mirror Earth. "We can measure as small as 0.5 meters per second. So, the only thing I can say is it's extremely expensive in terms of telescope time." The best use of this capability, he said, would be to follow up on Kepler candidates. Since Kepler's field of view is in the northern hemisphere, HARPS couldn't do that, but he had teamed up with Dimitar Sasselov, Dave Latham, and other exoplaneteers at Harvard to build a twin, HARPS North, which was nearing completion at the Italian Telescopio Nazionale Galileo, in the Canary Islands.)

The reconciliation of the different groups' statistics was hammered out in real time, as the meeting went on— something that doesn't happen often. In fact, said Marcy, "I've never seen anything like it. This is what happens when you run a meeting with three hundred smart people and deny them Internet in their rooms. They aren't holed up fiddling away; instead they're talking. Finding out that everybody's a human being and that we're much stronger when we work together. I was around here at about twelve thirty A.M. last night, and there were twenty-five people still there talking, with their laptops open. Yeah, okay, Kepler found five hundred planets, and the Swiss team also announced some planets, and there's a circumbinary planet [that was the Tatooine announcement]. Forget the circumbinary planet! Look at the sociological experiment that's been such an enormous success!" When Andrew Howard heard this sentiment, he smiled wryly and said, "Yeah, it's easy for Geoff to be so enthusiastic. *He* has Internet in his room."

For all the excitement about new planets and taking steps toward a reliable calculation of eta-sub-Earth, however, there was also an undercurrent of concern, especially for members of the Kepler team. Finding at least a handful of true Mirror Earths—the same size, the same density, the same temperature as the original—was always Kepler's goal. "That was the driver of the mission," Jack Lissauer said over lunch one day. But it was based on the assumption that the mission would last at least four years, and that the Sun was about typical in how noisy it is—how much it varies in brightness due to sunspots and flares and other activity.

Thanks to Kepler's unprecedented observations of Sun-like stars, however, the effort spearheaded by Ron Gilliland, the team now knew that the Sun is unusually quiet. Most Sun-like stars are about twice as noisy, which makes it just that much harder to tease out the signal of an Earth-size transiting planet. You can do it, but you need as many transits as possible, so the regular, repeating signal of a planet begins to stand out against the random flicker of noise. For an Earth-size planet in the habitable zone of a Sun-like star, transits come along only once a year. With a mission funded for only three and a half years, that's probably not enough to find a Mirror Earth. A few months after the Wyoming conference, the team would be petitioning NASA to extend the mission for another three and a half years. It's not as though the satellite would fall out of the sky, but it takes money to keep the analysis going.

   That's why Natalie Batalha was so worried about the upcoming budget review by NASA, which was now only four
months away. The Kepler team would do its best to convince
agency officials that they simply had to have funding for an
extended mission. But the Hubble Space Telescope was also
up for renewed funding. In a budget situation that appeared to
be getting worse by the month, there was a reasonable chance
that the most successful exoplanetology search in history
would be cut off before it could fulfill its mission. In the end,
NASA would announce in April 2012 that Kepler, Hubble,
and seven other missions under consideration would all be allowed to continue.

Chapter 18

SARA'S BIRTHDAY PARTY

T HE DAY AFTER the Boston meeting of the American Astronomical Society ended in May 2011, Sara Seager hosted a small and exclusive workshop at MIT to talk about the next forty years of exoplanet science. It was already clear to exoplaneteers that the wish list of telescopes and other instruments they'd been counting on to let them identify Mirror Earths and search these new worlds for life wasn't going to happen on any sort of reasonable schedule. In the face of that disappointment, how, Seager wanted to know, did her colleagues see the next four decades playing out?

The workshop took place on the top floor of a modern gem of a building at MIT that serves as headquarters for the Media Lab, an interdisciplinary laboratory that looks at how humans do and will interact with emerging and future technologies. As the guests trickled in and helped themselves to coffee, tea, and pastries, Seager urged everyone to take a quick stroll out onto an enormous balcony, with a wide view of the Charles River Basin and the Boston skyline beyond. And then, after a few minutes of milling around—pretty much everyone there knew everyone else, so there was more catching up than introductions—everyone moved into a small auditorium where Seager explained how things would work, and why she had convened the workshop.

She began by talking about cicadas, which are insects with a somewhat bizarre lifestyle. In the larval stage, cicadas lurk underground for years—thirteen in some places, seventeen in others—and then they emerge, all at once. They come out and feed on tree branches and leaves, and they hum. Millions upon millions of bugs fill the air with a loud, whirring noise that sounds like some sort of heavy machinery. It goes on for about three weeks. Then they mate and die. The new larvae burrow back into the ground, the bodies of millions of dead insects litter the ground, and all goes quiet for another thirteen or seventeen years. "I really want all of you to see that some day," she said.

The cicadas, Seager continued, "remind me a little bit of exoplanets, because, just like Earths, they may be rare, few, and far between but if you know where to look, they're everywhere." She recalled a conversation she and Paul Butler had had a few years earlier. "We were looking at a photograph of a cicada," she said, "and Paul told me, 'You know, the next time the cicadas return, we will be old.' And I went, 'Me, I'm never going to be old.' That's what I thought at the time. I was thirty-two, and I'd never thought about being old. TPF was always going to happen, I would always be young when TPF happened.' "

In just two months, Seager said, she'd be turning forty. "So I convened my friends here to come and tell me what they think should happen in the next forty years." The question was especially important to think about now, however, with TPF and SIM on the far back burner and the Webb telescope in financial trouble, Congress slashing budgets left and right, and word of Kepler's problem with stellar noise already spreading quietly around the exoplanetology community. She was also concerned that the field had lost some of the exciting, groundbreaking, out-on-the-edge feeling that Paul Butler and Geoff Marcy and Dave Charbonneau and Bill Borucki and the rest had brought to it. "I ask all of you to think," she said, "about how to continue the wave of exciting discoveries."

With that, the talks began. They were limited to ten minutes each, just enough time to convey a few big ideas in the most general sort of way, followed by questions. Dave Charbonneau got up to talk about the fact that a remarkable 246 M-dwarf stars lie within just ten parsecs, or thirty-two light-years, of Earth. They have distinct personalities, he said, "and so what I'd like to advocate for today is that we need to know and understand every single one of these stars." Based on the still-evolving eta-sub-Earth numbers coming from Kepler and from the radial-velocity surveys, he said, "the answer appears to be that we are guaranteed that there are bodies at the right distance from those M-stars [to be habitable]. We have no idea if they're inhabited but they're a good place to start looking."

Geoff Marcy was more downbeat: He talked about his frustration with all the canceled exoplanetology missions. Dave Latham talked about a new generation of gigantic ground-based telescopes—at a minimum, three times as big as the Keck, the largest in existence, with twenty-seven times the collecting area—armed with equally powerful spectroscopes that will pick up the radial-velocity signals of Mirror Earths by the dozens, perhaps as soon as the early 2020s. Matt Mountain, director of the Space Telescope Science Institute, made the case that the Webb telescope, along with a starshade, would be a more powerful planet-hunting telescope than anyone appreciated.

There were more than a dozen speakers in all, and it would be impossible to reproduce all of the ideas that were tossed around, especially in the Q&A sessions that followed the talks. One speaker made an especially powerful impression, however, and she wasn't an all-star of exoplanetology by a long shot. Her name was Rebecca Jensen-Clem; she was an MIT junior who had taken a class with Seager in which the students developed a satellite that was slated for a NASA launch sometime in 2012. It was a rectangular box, four inches by four inches by twelve inches, crammed with a small telescope and all sorts of electronics, and its job would be to look for transits—not around 150,000 stars, like Kepler, but around a single star. If it worked, this "ExoPlanetSat" could be mass-produced so quickly and cheaply that it could largely eliminate the need for full-blown space missions.

This wasn't a grander version of today's grandest technology: It was a complete rethinking of planet-hunting, and it was exactly this sort of innovation, Seager said during the Q&A, that could transform exoplanetology over the next forty years. "I hope," Jensen-Clem ended by saying, "that more students in the next forty years will have the opportunity to work with scientists and engineers as we did in ExoPlanetSat. I have every confidence that forty years from now, students will be able to come up to their professors and say, 'I don't remember a time when Earth was the only habitable planet.'"

And shortly after that, it was time to wrap things up. "Thank you," said Seager, taking the microphone one last time. "I'm going to finish with one remark and then a couple of other things. I want to tell you the origin of this picture." It was an image of two children pointing up and silhouetted against the glow of sunset, which itself was reflected in a pond. The children were hers—two boys, the older named Maxwell Solstice and the younger, Alexander Orion. "Does anybody recognize this setting?" she asked. "This is actually out at Walden Pond. It's almost impossible to take a picture of children, so you know how I actually got them to point?" Up in the darkening sky, she explained, outside the frame of the picture, was the crescent Moon, with a brilliant, sparkling gem—the planet Jupiter—sitting almost right between the points of the crescent. "They were just having a hysterical laugh because they're like, 'Oh, the Moon is eating Jupiter.'" Today, she said, pointing to one of the children, "that little one there, he's six years old and he walks around saying, 'I'm going to be an astronaut. I'm going to travel to the Earth-like planets that you find.'" And with that, the workshop was over. It was followed by a reception, and then a dinner in a conference room high up in the Green Building, to celebrate Seager's birthday.

Before everyone sat down to dinner, Seager had one last surprise. She led us to a tiny elevator where, five or six at a time, we rode up to the building's roof. It's the tallest building in Cambridge; the view was spectacular in all directions, as the exoplaneteers walked around, drinks in hand, catching up with one another and talking mostly about other things besides their research. Geoff Marcy decided to climb a ladder that led up into one of the white radar domes—the landmarks I used to see from down on the Charles Basin when I was a student. He couldn't get inside, which was probably just as well, but he did pose for some pictures.

The Charles River Basin itself was filled with sailboats on this warm, late-spring evening, the sailors unaware that the world's greatest exoplaneteers were looking down on them. Off in the west, the Sun was setting over Walden Pond, more or less—it was out in that direction somewhere. Although Seager had planned everything perfectly, she hadn't managed to re-create the Moon-Jupiter pairing that had fascinated her sons. Nobody seemed to mind. In small groups, everyone headed back downstairs, to toast Sara Seager's fortieth, and to talk, not only about what the program might be for a follow-up meeting forty years hence, but about the odds for finding a true Mirror Earth sometime in the next year. It had been just fifteen years since Geoff Marcy, holding forth at the far end of the dinner table, had been among the first scientists ever to detect an exoplanet—a giant, searingly hot ball of gas, where no living thing could possibly survive.

Now, thanks to the relentless efforts of Marcy and Charbonneau and Borucki and Batalha and dozens of other exoplaneteers around the world, scientists had cataloged hundreds upon hundreds of distant worlds, pushing to ever smaller and more hospitable and Earth-like planets all the time. They had found the first truly rocky planets. They knew that the Milky Way has more Neptune-size worlds than it has Jupiters, and more super-Earths than Neptunes, and almost certainly more Earths than super-Earths. They knew that planets are common, not just orbiting Sun-like stars, but orbiting the far more numerous red dwarf stars as well. They knew that with any sort of luck at all, the magic combination of factors—an Earth-size, rocky world, orbiting in the habitable zone of its parent star—would be found very, very soon. It might even be lurking in Kepler's vast data files, or in the data from observing runs by Michel Mayor's team, or in one of Marcy's data sets, just waiting to be analyzed and introduced to the world.

As they looked forward into the future from the vantage point of 2011, these superstars of exoplanetology knew that they couldn't really imagine where the field might stand forty years in the future. But they could easily see that a Mirror Earth might well be identified and confirmed within months rather than years. By the time Sara Seager turned forty-one, there was every chance that astronomers would be echoing my father's last words by proclaiming: "A Mirror Earth goes around a distant star—*that's* what's going on!"

ACKNOWLEDGMENTS

As is always the case with a work of nonfiction, my determination to write about exoplaneteers and their search for a Mirror Earth wouldn't have gone very far without the help of a long list of people. At the top is Cynthia Cannell, a terrific literary agent, but also a warm, thoughtful, and enormously supportive person. My editor, Jacqueline Johnson, has been a pleasure to work with. She didn't necessarily consider every word of my manuscript to be like a drop of precious gold, but that's a good thing: Like most writers, I need a smart, tasteful editor to help me weed out passages of writerly self-indulgence and keep the words flowing smoothly and clearly. I've had good editors and bad, and Jackie is one of the very best.

I obviously wouldn't have gotten very far without the help of dozens of exoplaneteers, including (in no particular order) Geoff Marcy, Debra Fischer, Natalie Batalha, Dave Latham, Sara Seager, Bill Borucki, Dave Charbonneau, Eric Ford, and many, many more. These scientists are among the busiest people I know, but they took the time to talk to me at great length, and in many cases more than once, to explain the work they were doing, and how and what they were finding, and why it mattered so much. Dan Fabrycky deserves special thanks: He read through the entire manuscript (before Jackie Johnson helped me work out the kinks!) to make sure I wouldn't make too many egregious errors.

I also want to thank my students at Princeton, Columbia, Johns Hopkins, and New York University, past and present, whose energy and intellectual curiosity invariably help shift my brain into high gear when it's idling unproductively in neutral. And I especially appreciate the willingness of two good friends and colleagues, Emily Elert and Alyson Kenward, who read parts of the manuscript and made helpful suggestions. Finally, as always, I thank my wife, Eileen, for being unfailingly supportive, even when this project intruded on family life. I owe her that trip to Vietnam, or India, or wherever she wants to go.

This book was supported in part by a generous grant from the Alfred P. Sloan Foundation.

NOTES

Chapter 1: THE MAN WHO LOOKED
FOR BLINKING STARS

13   "Astronomers have cracked the Milky Way like a piñata": Dennis Overbye, "Kepler Planet Hunter Finds
1,200 Possibilities," *New York Times*, February 2, 2011.

16   "Seth says 50 billion planets": Knight Science Journalism Tracker, February 20, 2011, http://ksjtracker.mit.
edu/2011/02/20/aaas-science-advisers-alert-keplers-new
-worlds-bilingual-brain-saving-solar-flare-batty-radar
-whos-covering-this-stuff/.

25   "Based on the stated assumptions": Borucki, W. J., and
Summers, A. L., "The Photometric Method of Detecting Other Planetary Systems," *Icarus* 58, no. 1 (1984):
121–134.

Chapter 2: THE MAN WHO LOOKED FOR WOBBLING STARS

32 "Marcy dragged himself out of bed": Michael D. Lemonick, *Other Worlds*, p. 63.

Chapter 3: HOT JUPITERS: WHO ORDERED THOSE?

51 "The companion is probably a brown dwarf": Latham, D., and Mazeh, T., et al., "The Unseen Companion of HD114762: A Probable Brown Dwarf," *Nature* 339 (1989): 38–40.

54 "We know that *stellar* companions": Struve, O., "Proposal for a Project of High-Precision Stellar Radial Velocity Work," *The Observatory* 72 (1952): 199–200.

Chapter 4: AN ANCIENT QUESTION

63 "there are infinite worlds": quoted in Steven J. Dick, *Plurality of Worlds*, p. 10.

64 "the worlds come into being as follows": quoted in Dick, *Plurality of Worlds*, p. 9.

68 "Rather than think that so many stars": quoted in Dick, *Plurality of Worlds*, p. 41.

69 He sneaked onto everyone's: I owe this insight, along with much of the information in this chapter, to the scholarly but highly readable histories written by Steven J. Dick, whose books are listed in the bibliography.

70   "There are countless suns and countless earths": quoted in Dick, *Plurality of Worlds*, p. 64.

72   "My Dear Kepler, what do you have to say about the principal philosophers": quoted in Dick, *Plurality of Worlds*, p. 118.

72   "the Moon is . . . inhabited": quoted in Dick, *Plurality of Worlds*, p. 19.

74   "Our Sun enlightens the Planets" quoted in Dick, *Plurality of Worlds*, p. 126.

75   "All the vast extent of the continents": quoted in Steven J. Dick, *The Biological Universe*, p. 69.

76   "Their singular aspect": quoted in Dick, *Biological Universe*, p. 70.

76   "Mr. Lowell went direct": Campbell, W. W., "Mars," *Science* 4, no. 86 (1896): 231–238.

78   "The probability of success": Cocconi, G., and Morrison, P., "Searching for Interstellar Communications," *Nature* 184, no. 4690 (1959): 844–846.

Chapter 9: WAITING FOR LAUNCH

149   "When I attended a conference": Jayawardhana, *Strange New Worlds*, p. 111.

151   "Sometime ago, R. W. Mandl": Einstein, A., "Lens-Like Action of a Star by the Deviation of Light in the Gravitational Field," *Science* 84, no. 2188 (1936): 506–507.

Chapter 10: KEPLER SCOOPED

176 "It is likely that this new world": Marcy, G., "Extrasolar Planets: Water World Larger than Earth," *Nature* 462, no. 17 (2009): 853–854.

Chapter 12: THE KEPLER ERA BEGINS

199 Fabrycky titled his creation the Kepler Orrery: The Kepler Orrery can be seen at www.ucolick.org/~fabrycky /kepler/.

Chapter 13: BEYOND KEPLER

208 "This is NASA's Hurricane Katrina": Kenneth Chang, "Telescope Is Behind Schedule and Over Budget, Panel Says," *New York Times*, November 10, 2010.

Chapter 14: HOW MANY EARTHS?

223 "Discussions of extrasolar planets": Kuchner, M., "Volatile-Rich Earth-Mass Planets in the Habitable Zone," *Astrophysical Journal* 596, no. 1 (2003): L105–108.

223 "A new family of planets": Léger, A., et al., "A New Family of Planets? 'Ocean-Planets,'" *Icarus* 169, no. 2 (2004): 499–504.

224 "What other possible kinds of planets": Kuchner, M., and Seager, S., "Extrasolar Carbon Planets," *Astrophysical Journal* (2005), http://arxiv.org/abs/astro-ph/0504214.

Chapter 17: ASTRONOMERS IN PARADISE

262 "When two elephants are waltzing": Michael D. Lem-
     onick, "Scientists Find a *Star Wars* World: One Planet,
     Two Suns," *Time*, September 16, 2011.

BIBLIOGRAPHY

Crowe, Michael J., *The Extraterrestrial Life Debate*, Cambridge, UK: Cambridge University Press, 1986.

Drake, Frank, and Sobel, Dava, *Is Anyone Out There? The Scientific Search for Extraterrestrial Intelligence*, New York: Delacorte Press, 1992.

Dick, Steven J., *The Biological Universe: The Twentieth Century Extraterrestrial Life Debate and the Limits of Science*, Cambridge, UK: Cambridge University Press, 1996.

———, *Plurality of Worlds: The Origins of the Extraterrestrial Life Debate from Democritus to Kant*, Cambridge, UK: Cambridge University Press, 1984

———, editor, *Many Worlds: The New Universe, Extraterrestrial Life & the Theological Implications*, Radnor, Penn.: Templeton Foundation Press, 2000.

Dick, Steven J., and Strick, James, *The Living Universe: NASA and the Development of Astrobiology*, New Brunswick, N.J.: Rutgers University Press, 2004.

Hoskin, Michael, *Discoverers of the Universe: William and Caroline Herschel*, Princeton, N.J.: Princeton University Press, 2011.

Impey, Chris, editor, *Talking About Life: Conversations on Astrobiology*, New York: Cambridge University Press, 2010.

Jayawardhana, Ray, *Strange New Worlds: The Search for Alien Planets and Life Beyond Our Solar System*, Princeton, N.J.: Princeton University Press, 2011.

Kasting, James, *How to Find a Habitable Planet*, Princeton, N.J.: Princeton University Press, 2010.

Lemonick, Michael D., *Other Worlds: The Search for Life in the Universe*, New York: Simon & Schuster, 1998.

Shostak, Seth, *Confessions of an Alien Hunter: A Scientist's Search for Extraterrestrial Intelligence*, Washington, D.C.: National Geographic Books, 2009.

INDEX